AF519846

T-Classes of Linear Estimators and the Theory of Successive Sampling

*This book is an edited version of the Lecture Notes on Advanced Theory of Sample Surveys of the author **Prof. B.D. Tikkiwal**. The lecture notes were out-growth of lecture material that evolved over many years of author's teaching and research work.*

G.C. Tikkiwal
Jai Narain Vyas University
Jodhpur

SCIENTIFIC
PUBLISHERS (INDIA)
www.scientificpub.com

Dr. G.C. Tikkiwal
Professor of Statistics
Dean, Faculty of Science
and
Head, Department of Mathematics & Statistics
Jai Narain Vyas University,
Jodhpur - 342 001, India
E-mail: gctikkiwal@yahoo.com

Published by:
Scientific Publishers (India)
5-A, New Pali Road, P.O. Box 91,
Jodhpur – 342 001 (India)

E-mail: info@scientificpub.com
www.scientificpub.com

© Scientific Publishers (India), 2010

Acknowledgement
The Publication of the work was financially supported by the University under the UGC Scheme. My thanks go the University Grants Commission and Jai Narain Vyas University for the publication grant. I also thank Dr. (Mrs.) Alka Khandelwal, the Women Scientist (A) under DST Scheme, Department of Mathematics and Statistics, Jai Narain Vyas University, Jodhpur, for the help in getting the manuscript in the print form. - ***G.C. Tikkiwal***

The responsibility for the facts stated, opinions expressed/and conclusions/ reaches is/ are entirely that of the editor and that the University accepts no responsibility for them.

ISBN: 978-81-7233-664-6

Lasertype set : Rajesh Ojha
Printed in India

Contents

Preface to First Write-up

The lecture notes,* on advanced theory of sample surveys, are out-growth of lecture material that evolved over many years of my teaching and research in this area. It was in the year 1959-60 that I included for the first time in my lectures, on Advanced Sample Surveys course, for M.A./M.Sc. Statistics students of Karnataka University, the material on T-classes largely based on Horvitz and Thompson's paper (1952). Since this paper had material only on sampling without replacement, I redefined T_2-class estimator in the manner given in Chapter 3 and asked the students to work out the details similar to those given by Horvitz and Thompson. Dr. Prabhu Ajgaonkar, then a Ph.D. research scholar working under my supervision in sampling on Successive Occasions, was also attending these lectures alongiwth the students. He came out with some results which were so encouraging that I asked him to switch over to the area of T-classes. The early results obtained in this area were reported in the Annals of Mathematical Statistics (Prabhu Ajgaonkar and Tikkiwal, 1961) in an abstract form. Some of these results were concerning T_2-class.

Tikkiwal (1965) reviewed the earlier work, broadly in the area of T-classes of Horvitz and Thompson (1952), Sen (1953), Yates and Grundy (1953), Prabhu Ajgaonkar and Tikkiwal (1961), Rao-Hartley-Cocharan (1962). He further defined in this paper extended T_2-class for multi-stage designs and after developing its general theory showed how the estimators due to Hansen and Hurwitz (1943), Horvitz and Thompson (1952), Sukhatme and Narain (1952), Durbin (1953), Ecimovic (1956) lie in the extended T_2-class. The results given in this paper provide most of the material of Chapter 3.

---

* A copy of the original lecture notes is available with the Library of International Statistical Institute at 428 Prinses Beatrixlaan, P.O. Box 950, 2270 A Z Voorburg, The Netherlands

The material in Chapter 3 alongwith the related material on sampling with varying probabilities in standard text books like that of Sukhatme and Sukhatme (1970) and the material on sampling theory, such as unbiased ratio method of estimation and non-sampling errors, as a left over material from one semester basic course on sample surveys, can form a one good semester course on Advance Theory of Sample Surveys. This observation is based on the experience of running this course in the University of Rajasthan for last four years under the semester system, introduced from 1972 and for couple of years under annual system.

The material in part (a) of Chapter 1 together with the two appendices at the end is meant to indicate how the first few lectures on a basic course in sample surveys should be organised to provide a sound base to this course. The material in parts (b) and (c) of Chapter 1 on the partition of sample space and then on the definition of T-classes of linear estimators and the technique of unordering and combined unordering (Tikkiwal, 1972), has been included, in much detail only this year, for the first time in my lectures on a semester course on sample surveys being given at the University of Rajasthan for some years. By the technique of unordering and combined unordering one can get all the results obtained otherwise by Basu (1958) and Pathak (1964) using the concept of sufficiency in survey sampling. This technique provides additional results in the process.

The material in Chapters 2, 4, 5 and 6 on the remaining T-classes along with the material on the theory of successive sampling in Chapter 7 in meant for one specialised sample survey course in a semester scheme. It should follow the one semester advanced course referred to above. The material in Chapter 7 on the theory of successive sampling should be supplemented from standard text books such as those by Sukhatme and Sukhatme and by Cochran. These text books deal with the theory under certain assumptions for an infinite population, where as this chapter deals with the theory under less restrictive assumptions for a finite population, there by making it possible to obtain the main results in these text books as a special case of the results presented in the chapter.

Some typed copies of all the above material except that given in Chapter 6 have been available in the Department of Statistics, University of Rajasthan for the use of concerned students and teachers. This material is presented in the various chapters with suitable modifications. The material in Chapter 6 on T_5, T_6 and T_7-classes has been added to the course from last year only. This material is based on the paper, by Bhargava and Tikkiwal (1973), which gives two new classes T_5 and T_6 based on the suitable partition of the sample space (Tikkiwal, 1972) used for developing the technique of unordering and combined unordering. The typed copies also contained the material on unified theory of sampling (Godambe, 1955) and on sufficiency in survey sampling. The latter material has been omitted in view of the addition of the material on unordering and combined unordering; and the former one has been abridged and improved upon in Chapter 6 in view of the fact that Godambe's general linear estimator is not general enough and is only a sub class of T_7 -class.

Because of my other pre-occupations at the time of writing these lecture notes and because of other environmental limitations there must have remained many weak points including mistakes of various types, omission etc. I shall be grateful to the Readers for pointing them out, so that I can improve upon these notes.

In preparing the above material, I have been ably assisted by Dr. C.L. Agarwal and Dr. G.C. Tikkiwal, the then two research fellows under my supervision. Dr. G.C., in fact, delivered lectures to M.A./M.Sc. Statistics students on the material in Sections 1 to 7 of Chapter 3, while it was being written. This arrangement, in which some one directly under the author rather than the author himself delivers the lecture worked much better in gauging the difficulties in presenting such material to the students. The stencilling etc. was done mainly by Mr. P.L. Sharma. My thanks go to all the above persons and to others who might have helped me in the task knowingly or unknowingly.

University of Rajasthan, **B.D. Tikkiwal**
Jaipur, India
November, 1977.

Preface to Second Write-up

Since bringing out these lecture notes in 1977, it was gradually felt, that to add to clarity the notations used by the author (1972) in describing the partition of sample space need to be improved upon. This has been done in Section 1.1 of Chapter 1 and consequential changes are made in subsequent five chapters on T-classes.

Another major change in this write up is to take away the material in part (a) of Chapter 1 together with the two appendices at the end of the first write up. This material was meant to indicate how a basic course in sample surveys should be organised. This material has now been considerably enlarged and appears separately (Author, 1985).

Soon after the first write up of these lecture notes, the paper by Bhargava (1978) appeared on the applications of the technique of unordering and combined unordering originally given by the author (1972). Some of these are now included in Chapter 2. In Chapter 1, an improved estimator for SRSWR is given using the technique of combined unordering. However, the estimator of the variance of the estimator based on distinct units in case of this sampling scheme is not included in these lecture notes, as it needs improvement.

From the lecture notes on the basic course, we know that all the estimators, in use, for equal probability design, lie in T_1-class. From Chapter 3 of these lecture notes, we know that most of the estimators, in use, for varying probability designs for uni-stage and multi-stage sampling, lie in author's T_2-class. Thus, the study of these two classes is a must for any one interested in the theory of sample surveys. The study of these classes as a whole is further necessiated by noting the fact that their theory (Tikkiwal, 1985) generates estimators for different sample designs. For example, if we are using Midzuno scheme of sampling with or without replacement, we get estimators for this sampling scheme in T_1-class, T_2-class and Godambe's general class discussed in Section 7 of Chapter 6. Some of these estimators can be further improved upon by the author's technique of unordering and combined unordering given in Chapter 7.

The theoretical treatment to the theory of successive sampling in Chapter 7 remains more or less the same as in the first write up. However, for the first time we find in a text book (Sukhatmes and Asok, 1984) treatment to the study of a finite population on successive occasions. But still the theoretical treatment in these lecture notes is far more exhaustive and complete.

My thanks go to the University Grants Commission and the University of Rajasthan for making it possible to organise in October-November, 1985 the All India Advanced Institute in Sample Surveys and Inference for training teachers in the new approach to the teaching of courses in sample surveys developed during the course of my teaching these courses for about three decades at the University of Rajasthan and my teaching at the University of Ilorin, Nigeria for about two and a half years as Visiting Professor there. As part of the programme of this Institute, these lectures were revised and up-dated.

I also thank Shri V.N. Rai and Dr. Diwakar Khare for their help in getting ready the manuscript in typed form and then photo-copied or stenciled. Lastly, but not the least, I thank Shri P.L. Sharma for his pains-taking efforts in typing or stenciling the manuscript.

B.D. Tikkiwal

University of Rajasthan,
Jaipur, India
July, 1986

CHAPTER 1

T-Classes of Linear Estimators and Techniques of Unordering and Combined Unordering

Part (a) : Definition of T-classes of Linear estimators

We present in this part first some preliminaries necessary for defining T-classes of linear estimators. The partition of sample space, given in Section 1.1, as a preliminary, was first presented by Tikkiwal (1972) while developing the theory of unordering and combined unordering given in part (b) of this chapter. The material in Sections 1.2 and 1.3 on the definition of T-classes is from the paper by Bhargava and Tikkiwal (1973, 78).

Definition of ordered set

Let S be a set of points x, y, z, ... Then S is an ordered set relative to $<$, if the following axioms hold :

(i) If x and y are disjoint points of S, then $x < y$ or $y < x$,

(ii) If $x < y$, then x and y are disjoint;

and

(iii) If $x < y$ and $y < z$, then $x < z$.

Definition of sample

A sample is an ordered set of certain sampling units from the population under study occurring at different draws.

Remark : It is understood that when the same unit occurs at two different draws, they will be regarded as disjoint for the purpose of defining a sample.

1.1 TIKKIWAL'S PARTITION OF SAMPLE SAPCE

We present in this section that partition of sample space (Tikkiwal, 1972) which will be used for (i) Defining T-classes of linear estimators in Section 1.2, (ii) Developing unified approach to unordering of estimators in Section 1.4 and, (iii) Developing technique of combiend unordering in Section 1.6. We first discuss partition of sample space with respect to a population of size 4.

Let the population be denoted by π : $(u_1, u_2, \ldots, u_N)$. Here, we shall replace the units, when necessary by their variate values, assuming of course for convenience that there is only one variate under study. Thus, let there be a population π : (x_1, x_2, x_3, x_4) of size four and let us draw a sample of size n = 3 from the population. Then all possible samples of size three are, in general, $(4)^3 = 64$ and are given below :

$x_1x_1x_1$	$x_1x_1x_2$	$x_1x_1x_3$	$x_1x_1x_4$	$x_2x_2x_3$	$x_2x_2x_4$
$x_2x_2x_2$	$x_1x_2x_1$	$x_1x_3x_1$	$x_1x_4x_1$	$x_2x_3x_2$	$x_2x_4x_2$
$x_3x_3x_3$	$x_2x_1x_1$	$x_3x_1x_1$	$x_4x_1x_1$	$x_3x_2x_2$	$x_4x_2x_2$
$x_4x_4x_4$	$x_2x_2x_1$	$x_3x_3x_1$	$x_4x_4x_1$	$x_3x_3x_2$	$x_4x_4x_2$
	$x_2x_1x_2$	$x_3x_1x_3$	$x_4x_1x_4$	$x_3x_2x_3$	$x_4x_2x_4$
	$x_1x_2x_2$	$x_1x_3x_3$	$x_1x_4x_4$	$x_2x_3x_3$	$x_2x_4x_4$
	$x_3x_3x_4$	$x_1x_2x_3$	$x_1x_2x_4$	$x_1x_3x_4$	$x_2x_3x_4$
	$x_3x_4x_3$	$x_1x_3x_2$	$x_1x_4x_2$	$x_1x_4x_3$	$x_2x_4x_3$
	$x_4x_3x_3$	$x_2x_1x_3$	$x_2x_1x_4$	$x_3x_1x_4$	$x_3x_2x_4$
	$x_4x_4x_3$	$x_2x_3x_1$	$x_2x_4x_1$	$x_3x_4x_1$	$x_3x_4x_2$
	$x_4x_3x_4$	$x_3x_1x_2$	$x_4x_1x_2$	$x_4x_1x_3$	$x_4x_2x_3$
	$x_3x_4x_4$	$x_3x_2x_1$	$x_4x_2x_1$	$x_4x_3x_1$	$x_4x_3x_2$

Thus the sample space Ω here consists of 64 sample points. Now we shall partition this sample space Ω at three stages as follows :

Let a sample point of Ω be denoted by s_t where t = 1, 2, . . ., 64. Let k be the number of distinct units in s_t. Then k may have three values i.e. either 1 or 2 or 3. We partition Ω at the first stage on the basis of the value of k; giving us three sub-sets Ω_1, Ω_2 and Ω_3 according to three values of k. The various sample points in different Ω_k (k = 1, 2, 3) are as follows :

$$\Omega_1 = \begin{bmatrix} x_1 & x_1 & x_1 \\ x_2 & x_2 & x_2 \\ x_3 & x_3 & x_3 \\ x_4 & x_4 & x_4 \end{bmatrix}$$

$$\Omega_2 = \begin{bmatrix} x_1x_1x_2 & x_1x_1x_3 & x_1x_1x_4 & x_2x_2x_3 & x_2x_2x_4 & x_3x_3x_4 \\ x_1x_2x_1 & x_1x_3x_1 & x_1x_4x_1 & x_2x_3x_2 & x_2x_4x_2 & x_3x_4x_3 \\ x_2x_1x_1 & x_3x_1x_1 & x_4x_1x_1 & x_3x_2x_2 & x_4x_2x_2 & x_4x_3x_3 \\ x_2x_2x_1 & x_3x_3x_1 & x_4x_4x_1 & x_3x_3x_2 & x_4x_4x_2 & x_4x_4x_3 \\ x_2x_1x_2 & x_3x_1x_3 & x_4x_1x_4 & x_3x_2x_3 & x_4x_2x_4 & x_4x_3x_4 \\ x_1x_2x_2 & x_1x_3x_3 & x_1x_4x_4 & x_2x_3x_3 & x_2x_4x_4 & x_3x_4x_4 \end{bmatrix}$$

$$\Omega_3 = \begin{bmatrix} x_1x_2x_3 & x_1x_2x_4 & x_1x_3x_4 & x_2x_3x_4 \\ x_1x_3x_2 & x_1x_4x_2 & x_1x_4x_3 & x_2x_4x_3 \\ x_2x_1x_3 & x_2x_1x_4 & x_3x_1x_4 & x_3x_2x_4 \\ x_2x_3x_1 & x_2x_4x_1 & x_3x_4x_1 & x_3x_4x_2 \\ x_3x_1x_2 & x_4x_1x_2 & x_4x_1x_3 & x_4x_2x_3 \\ x_3x_2x_1 & x_4x_2x_1 & x_4x_3x_1 & x_4x_3x_2 \end{bmatrix}$$

Therefore, Ω_1, Ω_2 and Ω_3 consist of 4, 36 and 24 sample points respectively. Since we can choose k (1 or 2 or 3) distinct units out of 4 in ${}^4C_k$ ways, thus, we again partition each Ω_k (k = 1, 2, 3) into ${}^4C_k$ ζ-classes. Thus; Ω_1 consists of ${}^4C_1 = 4$ ζ-classes in which each ζ-class consists of only one sample point. The Ω_2 consists of ${}^4C_2 = 6$ ζ-classes in which each ζ-class consists of six sample points. The Ω_3 consists of ${}^4C_3 = 4$ ζ-classes in which each ζ-class consists of six sample points. Therefore, Ω consists of 4 + 6 + 4 = 14 ζ-classes in all, which are given below :

$$\Omega_1 = \{\, \zeta_1 = x_1x_1x_1;\ \ \zeta_2 = x_2x_2x_2\ ;\ \zeta_3 = x_3x_3x_3\ ;\ \zeta_4 = x_4x_4x_4\ \};$$

$$\Omega_2 = \begin{bmatrix} \zeta_5 = \begin{bmatrix} x_1x_1x_2 \\ x_1x_2x_1 \\ x_2x_1x_1 \\ x_2x_2x_1 \\ x_2x_1x_2 \\ x_1x_2x_2 \end{bmatrix}; \zeta_6 = \begin{bmatrix} x_1x_1x_3 \\ x_1x_3x_1 \\ x_3x_1x_1 \\ x_3x_3x_1 \\ x_3x_1x_3 \\ x_1x_3x_3 \end{bmatrix}; \zeta_7 = \begin{bmatrix} x_1x_1x_4 \\ x_1x_4x_1 \\ x_4x_1x_1 \\ x_4x_4x_1 \\ x_4x_1x_4 \\ x_1x_4x_4 \end{bmatrix} \\ \zeta_8 = \begin{bmatrix} x_2x_2x_3 \\ x_2x_3x_2 \\ x_3x_2x_2 \\ x_3x_3x_2 \\ x_3x_2x_3 \\ x_2x_3x_3 \end{bmatrix}; \zeta_9 = \begin{bmatrix} x_2x_2x_4 \\ x_2x_4x_2 \\ x_4x_2x_2 \\ x_4x_4x_2 \\ x_4x_2x_4 \\ x_2x_4x_4 \end{bmatrix}; \zeta_{10} = \begin{bmatrix} x_3x_3x_4 \\ x_3x_4x_3 \\ x_4x_3x_3 \\ x_4x_4x_3 \\ x_4x_3x_4 \\ x_3x_4x_4 \end{bmatrix} \end{bmatrix}$$

and,

$$\Omega_3 = \left[\zeta_{11} = \begin{bmatrix} x_1x_2x_3 \\ x_1x_3x_2 \\ x_2x_1x_3 \\ x_2x_3x_1 \\ x_3x_1x_2 \\ x_3x_2x_1 \end{bmatrix}; \zeta_{12} = \begin{bmatrix} x_1x_2x_4 \\ x_1x_4x_2 \\ x_2x_1x_4 \\ x_2x_4x_1 \\ x_4x_1x_2 \\ x_4x_2x_1 \end{bmatrix}; \zeta_{13} = \begin{bmatrix} x_1x_3x_4 \\ x_1x_4x_3 \\ x_3x_1x_4 \\ x_3x_4x_1 \\ x_4x_1x_3 \\ x_4x_3x_1 \end{bmatrix}; \zeta_{14} = \begin{bmatrix} x_2x_3x_4 \\ x_2x_4x_3 \\ x_3x_2x_4 \\ x_3x_4x_2 \\ x_4x_2x_3 \\ x_4x_3x_2 \end{bmatrix} \right]$$

Let in a given ζ-class, a sample point s_t consist of i_j-th unit occurring R_{i_j} times such that $\sum_{j=1}^{k} R_{i_j} = 3$ for j = 1, 2, . . ., k and k = 1, 2, 3. Then ζ-class is further partitioned according to the different possible combinations of R_{i_j}'s in the given ζ-class. Such a type of partition of ζ-class, we call S-partition into sets of that ζ-class. The set itself is called the S-set. We observe that each ζ-class of set Ω_1 contains only one sample point in which the same unit is repeated three times. Therefore, no further partition is possible in these ζ-classes of sub-set Ω_1. We shall say that ζ-partition and S-partition coincide (or are same) in this case. Also in each of the ζ-classes of sub-set Ω_3, a sample point s_t contains three distinct units each occurring only once. Therefore, there is only one possible combination in this case and thus further partition is not possible in this case also. In case of the ζ-classes of sub-set Ω_2, i.e., when k = 2, there are two possible combinations ot two given distinct units i_1 and i_2 in a sample of three units : (i) i_1 unit occurs once and i_2 occurs twice and (ii) i_2 occurs once and i_1 occurs twice, for all $i_1 \neq i_2$. Thus, for any given k distinct units, taking ζ as consisting of as many sets S's as the number of possible vectors $R^{(i)} = (R_{i_1}, R_{i_2}, \ldots, R_{i_j})$ such that $\sum_{j=1}^{k} R_{i_j} =$ n, we rewrite Ω_1, Ω_2 and Ω_3 as follows :

$$\Omega_1 = \begin{bmatrix} \zeta_1 = S_1 = (x_1\,x_1\,x_1);\; \zeta_2 = S_2 = (x_2\,x_2\,x_2); \\ \zeta_3 = S_3 = (x_3\,x_3\,x_3);\; \zeta_4 = S_4 = (x_4\,x_4\,x_4) \end{bmatrix}$$

$$\Omega_2 = \begin{bmatrix} \zeta_5 = \begin{bmatrix} S_5 = \begin{bmatrix} x_1x_1x_2 \\ x_1x_2x_1 \\ x_2x_1x_1 \end{bmatrix} \\ S_6 = \begin{bmatrix} x_2x_2x_1 \\ x_2x_1x_2 \\ x_1x_2x_2 \end{bmatrix} \end{bmatrix} & \zeta_6 = \begin{bmatrix} S_7 = \begin{bmatrix} x_1x_1x_3 \\ x_1x_3x_1 \\ x_3x_1x_1 \end{bmatrix} \\ S_8 = \begin{bmatrix} x_3x_3x_1 \\ x_3x_1x_3 \\ x_1x_3x_3 \end{bmatrix} \end{bmatrix} & \zeta_7 = \begin{bmatrix} S_9 = \begin{bmatrix} x_1x_1x_4 \\ x_1x_4x_1 \\ x_4x_1x_1 \end{bmatrix} \\ S_{10} = \begin{bmatrix} x_4x_4x_1 \\ x_4x_1x_4 \\ x_1x_4x_4 \end{bmatrix} \end{bmatrix} \\ \zeta_8 = \begin{bmatrix} S_{11} = \begin{bmatrix} x_2x_2x_3 \\ x_2x_3x_2 \\ x_3x_2x_2 \end{bmatrix} \\ S_{12} = \begin{bmatrix} x_3x_3x_2 \\ x_3x_2x_3 \\ x_2x_3x_3 \end{bmatrix} \end{bmatrix} & \zeta_9 = \begin{bmatrix} S_{13} = \begin{bmatrix} x_2x_2x_4 \\ x_2x_4x_2 \\ x_4x_2x_2 \end{bmatrix} \\ S_{14} = \begin{bmatrix} x_4x_4x_2 \\ x_4x_2x_4 \\ x_2x_4x_4 \end{bmatrix} \end{bmatrix} & \zeta_{10} = \begin{bmatrix} S_{15} = \begin{bmatrix} x_3x_3x_4 \\ x_3x_4x_3 \\ x_4x_3x_3 \end{bmatrix} \\ S_{16} = \begin{bmatrix} x_4x_4x_3 \\ x_4x_3x_4 \\ x_3x_4x_4 \end{bmatrix} \end{bmatrix} \end{bmatrix}$$

$$\Omega_3 = \begin{bmatrix} \zeta_{11} = S_{17} = \begin{bmatrix} x_1x_2x_3 \\ x_1x_3x_2 \\ x_2x_1x_3 \\ x_2x_3x_1 \\ x_3x_1x_2 \\ x_3x_2x_1 \end{bmatrix} & \zeta_{12} = S_{18} = \begin{bmatrix} x_1x_2x_4 \\ x_1x_4x_2 \\ x_2x_1x_4 \\ x_2x_4x_1 \\ x_4x_1x_2 \\ x_4x_2x_1 \end{bmatrix} \\ \zeta_{13} = S_{19} = \begin{bmatrix} x_1x_3x_4 \\ x_1x_4x_3 \\ x_3x_1x_4 \\ x_3x_4x_1 \\ x_4x_1x_3 \\ x_4x_3x_1 \end{bmatrix} & \zeta_{14} = S_{20} = \begin{bmatrix} x_2x_3x_4 \\ x_2x_4x_3 \\ x_3x_1x_4 \\ x_3x_4x_2 \\ x_4x_2x_3 \\ x_4x_3x_2 \end{bmatrix} \end{bmatrix}$$

Now we deal with general case when population π consists of N units and sample size is n. The possible samples of size n are N^n, where $p(s_t) > 0$ and $\sum_{s_t} p(s_t) = 1$.

Let s_t-th sample, $t = 1, 2, \ldots, N^n$, consist of k distinct units for $k = 1, 2, \ldots, n$. Let i_j-th unit in the sample, $j = 1, 2, \ldots, k$, occur $R_{i_j}^{(p)} > 0$ times with $\sum_{j=1}^{k} R_{i_j}^{(p)} = n$. Let $R^{(p)} = R_{i_1}^{(p)}, R_{i_2}^{(p)}, \ldots, R_{i_j}^{(p)}$. For a given $R^{(p)}$, there are $M_{klp} = n! \Big/ \overset{k}{\underset{j=1}{\pi}} \left(R_{i_j}^{(p)}\right)!$ possible samples for different ordering of the same units. It may be noted that M_{klp} is independent of the nature of k distinct units. Let

$S_p\left\{\left(i_1, R_{i_1}^{(p)}\right), \ldots, \left(i_k, R_{i_k}^{(p)}\right)\right\}$ or S_p for brevity, be the set of $M_{kl}p$ samples, each sample having $(i_1, i_2, \ldots, i_k)$, k distinct units with i_j-th occurring $R_{i_j}^{(p)}$ times for j = 1, 2, . . . , k. Let M_{kl}, for a given set of k distinct units, be the possible number of ways, we can choose $R^{(p)}$ such that $\sum_{j=1}^{k} R_{i_j}^{(p)} = n$. That the number $M_{kl} = \binom{n-1}{k-1}$ is seen from the consideration that the problem of choosing vectors $R^{(p)}$ such that $\sum_{j=1}^{k} R_{i_j}^{(p)} = n$ amounts to filling k cells by n indistingui-shable objects so that no cell is empty. Let $\zeta_l(i_1, i_2, \ldots, i_k)$ or ζ_l for brevity, denote the class of these M_{kl} S-sets for a given set of $(i_1, i_2, \ldots, i_k)$, k distinct units. Since a set of k distinct units can be chosen in ${}^{N}C_k$ ways, so there are ${}^{N}C_k$ such classes. Thus there are, in all, $\sum_{k=1}^{n} {}^{N}C_k$ ζ-classes.

1.2 SET OF T-CLASSES OF LINEAR ESTIMATORS

A sample, which is an ordered set of certain sampling units, is generally obtained (or can be viewed as obtained) by drawing units one by one with or without replacement. In order to obtain linear function of the sample observations on the units drawn, the weights (or coefficients) in the linear function can depend on the draw, units drawn and the sample so obtained in the process. The weights can also depend upon the partition of the sample space. It is this nature of dependence of weights in the linear function which give rise to different T-classes of linear estimators as defined below.

(i) **T_1-class.** T_1-class of linear estimators is obtained by associating weight β_r to X_r, the observation on the unit drawn at r-th draw for r = 1, 2,. . ., n and then taking the sum. Thus,

$$\hat{T}_1 = \sum_{r=1}^{n} \beta_{ro}^{s_t} X_r$$

where, $\beta_{ro}^{s_t} = \beta_r$ for r = 1, 2, . . ., n and t = 1, 2, . . ., N^n

(ii) **T_2-class.** T_2-class of linear estimators is obtained by associating weight β_i to the observation x_i on the i-th unit of the population, i = 1, 2, . . ., N, if it occurs at the r-th draw for r=1, 2, . . ., n and then taking the sum over all n values. Thus

$$\hat{T}_2 = \sum_{r=1}^{n} \beta_{ro}^{s_t} X_r$$

where, $\beta_{ro}^{s_t} = \beta_i$ if $X_r = x_i$ for r = 1, 2, . . ., n and i = 1, 2, . . ., N in any given s_t.

(iii) **T_3-class.** T_3-class of linear estimators is obtained by associating weight β^{s_t} to the total of n observations for the s_t-th sample for t = 1, 2, . . ., N^n. Thus

$$\hat{T}_3 = \sum_{r=1}^{n} \beta_{ro}^{s_t} X_r$$

where, $\beta_{ro}^{s_t} = \beta^{s_t}$ for all r and t.

(iv) **T_4-class.** T_4-class of linear estimators is obtained by associating weight β_{ri} to the observation on the i-th unit drawn at the r-th draw for r = 1, 2, . . ., n and i = 1, 2, . . ., N and then taking the sum for all r. Thus,

$$\hat{T}_4 = \sum_{r=1}^{n} \beta_{ro}^{s_t} X_r$$

where, $\beta_{ro}^{s_t} = \beta_{ri}$ if $X_r = x_i$ for all r and i in given s_t.

(v) **T_5-class.** T_5-class of linear estimators is obtained from the given s_t-th sample lying in a particular set S_p by associating weight $\beta_{i_j}^{S_p}$ to the observation on the i_j-th unit for j = 1, 2, . . . , k; if it occurs at the r-th draw for r = 1, 2, . . . , n in the sample and then taking the sum. Thus,

$$\hat{T}_5 = \sum_{r=1}^{n} \beta_{ro}^{s_t} X_r$$

where, $\beta_{ro}^{s_t} = \beta_{i_j}^{S_p}$ if $X_r = x_{i_j}$ for r = 1, 2, . . ., n ; j = 1, 2, . . . , k and

$$s_t \in S_p \left\{ \left(i_1, R_{i_1}^{(p)}\right), \ldots, \left(i_k, R_{i_k}^{(p)}\right) \right\}$$

(vi) **T_6-class.** T_6-class of linear estimators is obtained from the given $s_t \varepsilon S_p \left\{ \left(i_1, R^{(p)}_{i_1} \right), \ldots, \left(i_k, R^{(p)}_{i_k} \right) \right\}$ by associating weight $\beta^{S_p}_r$ to the observation on a unit drawn at the r-th draw for r = 1, 2, . . ., n and then taking the sum. Thus,

$$\hat{T}_6 = \sum_{r=1}^{n} \beta^{s_t}_{ro} X_r$$

where, $\beta^{s_t}_{ro} = \beta^{Sp}_{r}$ for all $s_t \varepsilon S_p \left\{ \left(i_1, R^{(p)}_{i_1} \right), \ldots, \left(i_k, R^{(p)}_{i_k} \right) \right\}$

(vii) **T_7-class.** T_7-class of linear estimators is obtained by associating weight β^{st}_{ri} to the observation on the i-th unit in the s_t-th sample drawn at the r-th draw for t = 1, 2, . . ., N^n ; r = 1, 2, . . ., n and i = 1, 2, . . ., N and then taking the sum. Thus,

$$\hat{T}_7 = \sum_{r=1}^{n} \beta^{s_t}_{ro} X_r$$

where, $\beta^{st}_{ro} = \beta^{st}_{ri}$ if $X_r = x_i$ for all r and i in a given s_t.

Now, we give below the various linear estimators for different T-classes for the samples of ζ_5-class, viz.

$$\zeta_5 = \left[S_5 = \begin{bmatrix} s_1 = (x_1\, x_1\, x_2) \\ s_2 = (x_1\, x_2\, x_1) \\ s_3 = (x_2\, x_1\, x_1) \end{bmatrix} \text{ and } S_6 = \begin{bmatrix} s_4 = (x_2\, x_2\, x_1) \\ s_5 = (x_2\, x_1\, x_2) \\ s_6 = (x_1\, x_2\, x_2) \end{bmatrix} \right]$$

given in Section 1.1.

(i) Construction of T_1 - class of linear estimators

Sample No.	Estimator
1	$\beta_1 x_1 + \beta_2 x_1 + \beta_3 x_2$
2	$\beta_1 x_1 + \beta_2 x_2 + \beta_3 x_1$
3	$\beta_1 x_2 + \beta_2 x_1 + \beta_3 x_1$
4	$\beta_1 x_2 + \beta_2 x_2 + \beta_3 x_1$
5	$\beta_1 x_2 + \beta_2 x_1 + \beta_3 x_2$
6	$\beta_1 x_1 + \beta_2 x_2 + \beta_3 x_2$

(ii) Construction of T_2-class of linear estimators

Sample No.	Estimator
1	$\beta_1 x_1 + \beta_1 x_1 + \beta_2 x_2$
2	$\beta_1 x_1 + \beta_1 x_1 + \beta_2 x_2$
3	$\beta_1 x_1 + \beta_1 x_1 + \beta_2 x_2$
4	$\beta_2 x_2 + \beta_2 x_2 + \beta_1 x_1$
5	$\beta_2 x_2 + \beta_2 x_2 + \beta_1 x_1$
6	$\beta_2 x_2 + \beta_2 x_2 + \beta_1 x_1$

(iii) Construction of T_3-class of linear estimators

Sample No.	Estimator
1	$\beta^{s_1} (x_1 + x_1 + x_2)$
2	$\beta^{s_2} (x_1 + x_2 + x_1)$
3	$\beta^{s_3} (x_2 + x_1 + x_1)$
4	$\beta^{s_4} (x_2 + x_2 + x_1)$
5	$\beta^{s_5} (x_2 + x_1 + x_2)$
6	$\beta^{s_6} (x_1 + x_2 + x_2)$

(iv) Construction of T_4-class of linear estimators

Sample No.	Estimator
1	$\beta_{11} x_1 + \beta_{21} x_1 + \beta_{32} x_2$
2	$\beta_{11} x_1 + \beta_{22} x_2 + \beta_{31} x_1$
3	$\beta_{12} x_2 + \beta_{21} x_1 + \beta_{31} x_1$
4	$\beta_{12} x_2 + \beta_{22} x_2 + \beta_{31} x_1$
5	$\beta_{12} x_2 + \beta_{21} x_1 + \beta_{32} x_2$
6	$\beta_{11} x_1 + \beta_{22} x_2 + \beta_{32} x_2$

(v) Construction of T_5-class of linear estimators

Sample No.	Estimator
1	$\beta_1^{S5} x_1 + \beta_1^{S5} x_1 + \beta_2^{S5} x_2$
2	$\beta_1^{S5} x_1 + \beta_1^{S5} x_1 + \beta_2^{S5} x_2$
3	$\beta_1^{S5} x_1 + \beta_1^{S5} x_1 + \beta_2^{S5} x_2$
4	$\beta_2^{S6} x_2 + \beta_2^{S6} x_2 + \beta_1^{S6} x_1$
5	$\beta_2^{S6} x_2 + \beta_2^{S6} x_2 + \beta_1^{S6} x_1$
6	$\beta_2^{S6} x_2 + \beta_2^{S6} x_2 + \beta_1^{S6} x_1$

(vi) Construction of T_6-class of linear estimators

Sample No.	Estimator
1	$\beta_1^{S5} x_1 + \beta_2^{S5} x_1 + \beta_3^{S5} x_2$
2	$\beta_1^{S5} x_1 + \beta_2^{S5} x_2 + \beta_3^{S5} x_1$
3	$\beta_1^{S5} x_2 + \beta_2^{S5} x_1 + \beta_3^{S5} x_1$
4	$\beta_1^{S6} x_2 + \beta_2^{S6} x_2 + \beta_3^{S6} x_1$
5	$\beta_1^{S6} x_2 + \beta_2^{S6} x_1 + \beta_3^{S6} x_2$
6	$\beta_1^{S6} x_1 + \beta_2^{S6} x_2 + \beta_3^{S6} x_2$

(vii) Construction of T_7-class of linear estimators

Sample No.	Estimator
1	$\beta_{11}^{s_1} x_1 + \beta_{21}^{s_1} x_1 + \beta_{32}^{s_1} x_2$
2	$\beta_{11}^{s_2} x_1 + \beta_{22}^{s_2} x_1 + \beta_{31}^{s_2} x_1$
3	$\beta_{12}^{s_3} x_2 + \beta_{21}^{s_3} x_1 + \beta_{31}^{s_3} x_1$
4	$\beta_{12}^{s_4} x_2 + \beta_{22}^{s_4} x_2 + \beta_{31}^{s_4} x_1$
5	$\beta_{12}^{s_5} x_2 + \beta_{21}^{s_5} x_1 + \beta_{32}^{s_5} x_2$
6	$\beta_{11}^{s_6} x_1 + \beta_{22}^{s_6} x_2 + \beta_{32}^{s_6} x_2$

1.3 WIDENESS OF THE SET OF SEVEN CLASSES OF LINEAR ESTIMATORS

Hansen and Hurwitz (1943, 1949) first suggested the use of sampling with varying probabilities. Narain (1951) on one hand and Horvitz and Thompson (1952) on the other, apparently working independently, developed the theory further for sampling with varying probabilities and without replacement. Though, the main estimator and its variancc, given by Horvitz and Thompson in their Eqs. (6) and (9), can be obtained with slight modification in the estimator and its variance given by Narain, the two authors together presented the theory in a way suggestive of further extension. They noted that there can be a number of classes of linear estimators and mentioned three such estimators by way of illustration. They confined their detailed discussions to sampling with varying probabilities and without replacement. Tikkiwal (1960), Prabhu Ajgaonkar and Tikkiwal (1961) and Prabhu Ajgaonkar (1962) gave a unified theory for sampling with varying probabilities with or without replacement. Koop (1961) and Godambe (1955) also dealt with T-classes both extensively and intensively.

The set of seven classes of linear estimator as discussed in Section 1.2 has been developed by Tikkiwal with the assistance of two research scholars, Prabhu Ajagaonkar and Bhargava while working for their Ph.D. at different times. These up-dated definitions are presented by Bhargava and Tikkiwal (1973). It is to be noted that T_1, T_4 and T_7 classes of linear estimators given here are the same as given by others. The T_6-class of Koop corresponds to T_3-class of Section 1.2. The T_5 and T_6-classes have been redefined by Bhargava and Tikkiwal (1973). Now we shall show that the T_2, T_3 and T_5-classes due to Koop, Godambe or both and the general class due to Godambe, for sampling with or without replacement, are sub-classes of some classes given in Section 1.2. The proof is contained in the following Theorem.

Theorem 1.3.1 (Bhargava and Tikkiwal, 1978)

(a) An estimator in any one of the following classes lies in T_5-class given in Section 1.2: (i) $T_{2(Gk)}$- class due to Godambe and Koop, where $\hat{T}_{2(Gk)} = \sum_{i \in s_t} \beta_i x_i$; (ii) $T_{3(k)}$ -class due to Koop, where

$\hat{T}_{3(k)} = \beta^{\zeta_l} \sum_{i \in \zeta_l} x_i, l = 1, 2, \ldots, {}^N C_n$ for SWOR and $l = 1, 2, \ldots, \sum_{k=1}^{n} {}^N C_k$ for SWR and (iii) $T_{5(k)}$-class due to Koop, where $\hat{T}_{5(k)} = \sum_{i \in \zeta_l} \beta_i^{\zeta_l} x_i$.

(b) An estimator in T_3-class or general class, where $\hat{T}_3 = \beta^{s_t} \sum_{i \in s_t} x_i$ and $\hat{T}_g = \sum_{i \in s_t} \beta_i^{s_t} x_i$ denote the linear estimators of T_3-class and general class respectively, both classes due to Godambe, lie in T_7-class given above.

Proof : (a) Let $\hat{T}_5 = \sum_{j=1}^{k} x_{i_j} R_{i_j}^{(p)} \beta_{i_j}^{S_p}$ be an estimator in T_5-class for all $s_t \in S_p \left\{\left(i_1, R_{i_1}^{(p)}\right), \ldots, \left(i_k, R_{i_k}^{(p)}\right)\right\}$. Let $\hat{T}_{2(Gk)} = \sum_{j=1}^{k} (x_{i_j} \beta_{i_j})$ be any one given estimator in $T_{2(Gk)}$-class due to Godambe and Koop for all $s_t \in S_p \left\{\left(i_1, R_{i_1}^{(p)}\right), \ldots, \left(i_k, R_{i_k}^{(p)}\right)\right\}$ and for $S_p \left\{\left(i_1, R_{i_1}^{(p)}\right), \ldots, \left(i_k, R_{i_k}^{(p)}\right)\right\} \in \zeta_l \left(i_1, i_2, \ldots, i_k\right)$. This estimator can be obtained from $\hat{T}_5$ by taking $\beta_{i_j}^{S_p} = \beta_{i_j} / R_{i_j}^{(p)}$ for j = 1, 2, . . ., k. Similarly an estimator in $T_{3(k)}$-class given by $\beta^{\zeta_l} \left(\sum_{j=1}^{k} x_{i_j} \right)$ and an estimator in T_5 – class given by $\sum_{j=1}^{k} \left(x_{i_j} \beta_{i_j}^{\zeta_l} \right)$, both based on an $s_t \in S_p \left\{\left(i_1, R_{i_1}^{(p)}\right), \ldots, \left(i_k, R_{i_k}^{(p)}\right)\right\} \in \zeta_l \left(i_1, i_2, \ldots, i_k\right)$ can be obtained from $\hat{T}_5$ by taking $\beta_{i_j}^{S_p} = \beta^{\zeta_l} / R_{i_j}^{(p)}$ and $\beta_{i_j}^{S_p} = \beta_{i_j}^{\zeta_l} / R_{i_j}^{(p)}$ respectively.

(b) Let $\beta^{s_t} \left(\sum_{j=1}^{k} x_{i_j} \right)$ and $\sum_{j=1}^{k} x_{i_j} \beta_{i_j}^{s_t}$ be two estimators, former one in T_3-class and latter one in general class for an $s_t \in S_p \left\{\left(i_1, R_{i_1}^{(p)}\right), \ldots, \left(i_k, R_{i_k}^{(p)}\right)\right\}$. Both these estimators can be obtained from the estimator in T_7-class, as defined in Section 1.2, by taking $\beta_{ro}^{s_t} = \beta^{s_t} / R_{i_j}^{(p)}$ in the first case and $\beta_{ro}^{s_t} = \beta_{i_j}^{s_t} / R_{i_j}^{(p)}$ in the second case, if $X_r = x_{i_j}$.

Part (b) : Techniques of unordering and combined unordering

Introduction

For samping with varying probabilities without replacement, Murthy (1957) gave a technique of unordering an ordered estimator based on an ordered sample s_t to get an improved estimator, in the sense of efficiency, over each of that based on different ordered samples s_t containing the same set of units. Tikkiwal (1972) gave a unified approach, based on the partition of sample space given in Section 1.2, for unordering an ordered estimator when there is sampling with or without replacement from, which Murthy's results follow as a special case. For sampling with replacement, he further gave the technique of combined unordering of an unordered estimator, resulting in further improvement of efficiency. Bhargava (1978) used the technique of combined unordering to obtain new results in addition to those given by Basu (1958) and Pathak (1966) by a different method based on the concept of sufficiency in survey sampling. We present in part (b) here some of the above salient results.

1.4 UNORDERING OF ESTIMATORS FOR SAMPLING WITH VARYING PROBABILITIES WITH OR WITHOUT REPLACEMENT - A UNIFIED APPROACH

In this section, we present the unified approach for unordering an ordered estimator for sampling with varying probabilities with or without replacement.

Let $\hat{T}_{s_t}$ be an estimator of the population total T based on t-th sample (an ordered set by definition) consisting of k distint units : $(i_1, i_2, \ldots, i_k)$, for $t = 1, 2, \ldots, N^n$. $\hat{T}_{s_t}$, be definition is said to be ordered estimator and, in general, assumes different values for different samples, in a given S_p consisting of the same k distinct units. One may say that $\hat{T}_{s_t}$ is an unordered estimator, if $\hat{T}_{s_t} = C$, a constant quantity, for all $s_t \in S_p$. The following results in Theorem 1.4.1. hold good whether $\hat{T}_{s_t}$ is an ordered or unordered estimator. Let

$$\hat{T}_{uo}^{(p)} = \sum_{t=1}^{M_{k/p}} \hat{T}_{s_t}\, p(s_t) \Big/ \sum_{t=1}^{M_{k/p}} p(s_t) \tag{1.4.1}$$

It may be noted that the probability $P(S_p)$ of the set S_p under consideration is equal to $\sum_{t=1}^{M_{k/p}} p(s_t)$. Now, we have the following theorem :

Theorem 1.4.1

(i) $E\left(\hat{T}_{uo}^{(p)}\right) = E\left(\hat{T}_{s_t}\right)$ (1.4.2)

(ii) $\mathrm{Var}\left(\hat{T}_{uo}^{(p)}\right) \leq \mathrm{Var}\left(\hat{T}_{s_t}\right)$ (1.4.3)

This theorem would mean that if $\hat{T}_{s_t}$ is an unbiased estimator for population total T, then the estimator $\hat{T}_{uo}^{(p)}$, obtained by combinig different $\hat{T}_{s_t}$ based on the samples $s_t \in S_p$, in the way done in Eq.(1.4.1.), is also unbiased for T and its variance is less than or equal to that of each of the ordered estimator $\hat{T}_{s_t}$. $\hat{T}_{uo}^{(p)}$ is said to be the unordered estimator for the population total T. We shall, hence onward, assume that $\hat{T}_{s_t}$ is, in fact, an unbiased estimator of T.

Proof :

(i)
$$E\left(\hat{T}_{uo}^{(p)}\right) = \sum_{k=1}^{n} \sum_{l=1}^{\binom{N}{k}} \sum_{p=1}^{M_{kl}} \left(\hat{T}_{uo}^{(p)}\right) P(S_p)$$

$$= \sum_{k=1}^{n} \sum_{l=1}^{\binom{N}{k}} \sum_{p=1}^{M_{kl}} \sum_{t=1}^{M_{k/p}} \left(\hat{T}_{s_t}\right) p(s_t)$$

$$= E\left(\hat{T}_{s_t}\right) = T$$

(ii)
$$\mathrm{Var}\left(\hat{T}_{uo}^{(p)}\right) = E\left(\hat{T}_{uo}^{(p)}\right)^2 - \left\{E\left(\hat{T}_{uo}^{(p)}\right)\right\}^2$$

$$= E\left(\hat{T}_{uo}^{(p)}\right)^2 - T^2$$

$$= \sum_{k=1}^{n} \sum_{l=1}^{\binom{N}{k}} \sum_{p=1}^{M_{kl}} \left\{\frac{\sum_{t=1}^{M_{k/p}} \hat{T}_{s_t} p(s_t)}{P(S_p)}\right\}^2 P(S_p) - T^2$$

$$= \sum_{k=1}^{n} \sum_{l=1}^{\binom{N}{k}} \sum_{p=1}^{M_{kl}} \left\{ \sum_{t=1}^{M_{klp}} \hat{T}_{s_t}\, p(s_t) \right\}^2 / P(S_p) - T^2$$

Similarly,

$$\mathrm{Var}\left(\hat{T}_{s_t}\right) = \sum_{k=1}^{n} \sum_{l=1}^{\binom{N}{k}} \sum_{p=1}^{M_{kl}} \sum_{t=1}^{M_{klp}} \hat{T}^2_{s_t}\, p(s_t) - T^2$$

Therefore,

$$\mathrm{Var}\left(\hat{T}_{s_t}\right) - \mathrm{Var}\left(\hat{T}^{(p)}_{uo}\right) = \sum_{k=1}^{n} \sum_{l=1}^{\binom{N}{k}} \sum_{p=1}^{M_{kl}} \sum_{t=1}^{M_{klp}} \left(\hat{T}_{s_t} - \hat{T}^{(p)}_{uo}\right)^2 p(s_t) \geq 0$$

This completes the proof of the Theorem.

Now we state and prove the following two Corollaries for this unified approach analogous to those given earlier by Murthy (1957) for sampling without replacement.

Corollary 1.4.1

Let $\hat{V}_{s_t}$ be an ordered linear estimator of Var $\left(\hat{T}_{s_t}\right)$, then the corresponding unordered estimator $\hat{V}^{(p)}_{uo}$ is given by

$$\hat{V}^{(p)}_{uo} = \sum_{t=1}^{M_{klp}} \hat{V}_{s_t}\, p'(s_t)$$

where, $p'(s_t) = p(s_t) / \sum_{t=1}^{M_{klp}} p(s_t)$

The unordered estimator is an improved estimator of the same parameter Var $\left(\hat{T}_{s_t}\right)$, which is at least as efficient as any of the ordered estimator $\hat{V}_{s_t}$ based on any given $s_t \in S_p$. If $\hat{V}_{s_t}$ be an unbiased estimator of Var $\left(\hat{T}_{s_t}\right)$, then $\hat{V}^{(p)}_{uo}$ is also unbiased estimator of Var $\left(\hat{T}_{s_t}\right)$.

The proof of the Corollary is on the same lines as that of Theorem 1.4.1.

Remark 1.4.1

This Corollary and the original Theorem can be put in slightly more general form by considering second order moments of the estimators about an arbitrary parameter, not necessarily equal to their expectations.

Proof :

The remark about the first and second parts of the Theorem and hence Corollary 1.4.1 follow by noting the following :

(i) $$E\left(\hat{T}_{s_t} - \theta\right)^2 = \text{Var}\left(\hat{T}_{s_t}\right) + (T - \theta)^2$$ and

(ii) $$E\left(\hat{T}_{uo}^{(p)} - \theta\right)^2 = \text{Var}\left(\hat{T}_{uo}^{(p)}\right) + (T - \theta)^2$$

For example the modified second part of Theorem 1.4.1 is $E\left(\hat{T}_{uo}^{(p)} - \theta\right)^2 \leq E\left(\hat{T}_{s_t} - \theta\right)^2$ which is true if $\text{Var}\left(\hat{T}_{uo}^{(p)}\right) \leq \text{Var}\left(\hat{T}_{s_t}\right)$. Since the latter is shown to be true, the modified part is also true.

Corollary 1.4.2

An unbiased linear estimator $\hat{V}_{uo}^{(p)}$ of $\text{Var}\left(\hat{T}_{uo}^{(p)}\right)$ is given by

$$\hat{V}_{uo}^{(p)} = \left(\hat{T}_{uo}^{(p)}\right)^2 - \sum_{t=1}^{M_{k/p}} \left(\hat{T}_{s_t}^2 - \hat{V}_{s_t}\right) p'(s_t) \tag{1.4.4}$$

Provided $\hat{V}_{s_t}$ is an unbiased estimator of $\text{Var}\left(\hat{T}_{s_t}\right)$.

Proof :

$$E\left(\hat{V}_{uo}^{(p)}\right) = E\left(\hat{T}_{uo}^{(p)}\right)^2 - E\left\{\sum_{t=1}^{M_{k/p}} \left(\hat{T}_{s_t}\right)^2 p'(s_t)\right\} + \left\{\sum_{t=1}^{M_{k/p}} \left(\hat{V}_{s_t}\right) p'(s_t)\right\}$$

$$= E\left(\hat{T}_{uo}^{(p)}\right)^2 - E\left(\hat{T}_{s_t}^2\right) + V\left(\hat{T}_{s_t}\right)$$

$$= E\left(\hat{T}_{uo}^{(p)}\right)^2 - T^2$$

as, $V\left(\hat{T}_{s_t}\right) = E\left(\hat{T}_{s_t}^2\right) - T^2$.

Therefore, $E\left(\hat{V}_{uo}^{(p)}\right) = V\left(\hat{T}_{uo}^{(p)}\right)$.

This completes the proof.

1.5 MURTHY'S RESULTS OF UNORDERING OF ORDERED ESTIMATORS FOR SAMPLING WITH VARYING PROBABILITIES WITHOUT REPLACEMENT

In Section 1.4, the unified approach has been given for unordering an ordered estimator when there is sampling with varying probabilities with or without replacement. The various results due to Murthy (1957) for sampling with varying probabilities without replacement can be obtained by putting $k = n$, $M_{kl} = 1, M_{kl p} = n!$ at appropriate places in Section 1.4. For example Eq. (1.4.1), in this manner, will give

$$\hat{T}_{uo}^{(p)} = \sum_{t=1}^{n!} \hat{T}_{s_t}\, p'(s_t) \tag{1.5.1}$$

where, $p'(s_t) = p(s_t) / \sum_{t=1}^{n!} p(s_t)$ a result due to Murthy (1957).

Tikkiwal (1972), using the partition of sample space further, extended this technique of unordering to get still an improved estimator over each of the unordered estimator $\hat{T}_{uo}^{(p)}$ based on $M_{kl p}$ samples. He showed that the estimation procedure, in which we further combine the $M_{kl p}$ unordered estimators $\hat{T}_{uo}^{(p)}$, provides further improvement.

He called this technique of further combining the unordered estimators as combined unordering and the estimator resulted therefrom as combined unordered estimator $\hat{T}_{cu}^{(l)}$. For sampling without replacement, where $M_{kl} = 1$, there is no combined unordering except in a trivial sense. But, for sampling with replacement, where $M_{kl} > 1$; this technique can be used profitably.

It has been shown by Tikkiwal, B.D. and Tikkiwal, G.C. (1979) that this technique is more serviceable than the known alternative technique advanced by the various authors, particularly those by Godambe and Basu during last two decades or so. Because of the importance of this technique, we present the same in detail in the next section.

1.6. THE TECHNIQUE OF COMBINED UNORDERING

Let the combined unordered estimator $\hat{T}_{cu}^{(l)}$, based on M_{kl} unordered estimators $\hat{T}_{uo}^{(p)}$, be given by

$$\hat{T}_{cu}^{(l)} = \sum_{p=1}^{M_{kl}} P(S_p)\ \hat{T}_{uo}^{(p)} \Big/ \sum_{p=1}^{M_{kl}} P(S_p) \tag{1.6.1}$$

We state below the following Theorems and its Corollaries with proofs, their proofs being on the lines of the proof of Theorem 1.4.1 and its Corollaries. The various Remarks of Section 1.4 still hold good here as well with necessary modifications.

Theorem 1.6.1

$$\text{(i)}\quad E\left(\hat{T}_{cu}^{(l)}\right) = E\left(\hat{T}_{uo}^{(p)}\right) \tag{1.6.2}$$

and $$\text{(ii)}\quad \operatorname{Var}\left(\hat{T}_{cu}^{(l)}\right) \leq \operatorname{Var}\left(\hat{T}_{uo}^{(p)}\right) \tag{1.6.3}$$

Proof :

$$\text{(i)}\quad E\left(\hat{T}_{cu}^{(l)}\right) = \sum_{k=1}^{n} \sum_{l=1}^{\binom{N}{k}} \hat{T}_{cu}^{(l)} \sum_{p=1}^{M_{kl}} P(S_p)$$

$$= \sum_{k=1}^{n} \sum_{l=1}^{\binom{N}{k}} \sum_{p=1}^{M_{kl}} \hat{T}_{uo}^{(p)}\ P(S_p)$$

$$= E\left(\hat{T}_{uo}^{(p)}\right) = T$$

We observe that if $\hat{T}_{uo}^{(p)}$ is unbiased estimator of population total T, then $\hat{T}_{cu}^{(l)}$ is also unbiased estimator of T.

$$\text{(ii)}\quad \operatorname{Var}\left(\hat{T}_{cu}^{(l)}\right) = E\left(T_{cu}^{(l)}\right)^2 - T^2$$

$$= \sum_{k=1}^{n} \sum_{l=1}^{\binom{N}{k}} \left(\hat{T}_{cu}^{(l)}\right)^2 \sum_{p=1}^{M_{kl}} P(S_p) - T^2 \tag{1.6.4}$$

and,

$$\operatorname{Var}\left(\hat{T}_{uo}^{(p)}\right) = E\left(\hat{T}_{uo}^{(p)}\right)^2 - T^2$$

$$= \sum_{k=1}^{n} \sum_{l=1}^{\binom{N}{k}} \sum_{p=1}^{M_{kl}} \left(\hat{T}_{uo}^{(p)}\right)^2 P(S_p) - T^2 \qquad (1.6.5)$$

Therefore,

$$\operatorname{Var}\left(\hat{T}_{uo}^{(p)}\right) - \operatorname{Var}\left(\hat{T}_{cu}^{(l)}\right) = \sum_{k=1}^{n} \sum_{l=1}^{\binom{N}{k}} \sum_{p=1}^{M_{kl}} \left(\hat{T}_{uo}^{(p)}\right)^2 P(S_p)$$

$$- \sum_{k=1}^{n} \sum_{l=1}^{\binom{N}{k}} \sum_{p=1}^{M_{kl}} \left(T_{cu}^{(l)}\right)^2 P(S_p)$$

$$= \sum_{k=1}^{n} \sum_{l=1}^{\binom{N}{k}} \sum_{p=1}^{M_{kl}} \left(\hat{T}_{uo}^{(p)} - \hat{T}_{cu}^{(l)}\right)^2 P(S_p) \geq 0$$

So $\operatorname{Var}\left(\hat{T}_{uo}^{(p)}\right) \geq \operatorname{Var}\left(\hat{T}_{cu}^{(l)}\right)$.

Hence, the Theorem 1.6.1.

Corollary 1.6.1

Let $\hat{V}_{cu}^{(l)} = \sum_{p=1}^{M_{kl}} P'(S_p)\, \hat{V}_{uo}^{(p)}$, be a combined unordered estimator of $\operatorname{Var}\left(\hat{T}_{uo}^{(p)}\right)$, where $P'(S_p) = P(S_p) / \sum_{p=1}^{M_{kl}} P(S_p)$. Then $\hat{V}_{cu}^{(l)}$ is an estimator which is at least as efficient as any of the unordered estimator $\hat{V}_{uo}^{(p)}$ given by Eq. (1.4.4) for p = 1, 2, . . ., M_{kl} . Also if, $\hat{V}_{uo}^{(p)}$ is unbiased, then $\hat{V}_{cu}^{(l)}$ is also unbiased for $\operatorname{Var}\left(\hat{T}_{uo}^{(p)}\right)$.

The proof of this Corollary is on the same lines as that of Theorem 1.6.1.

Corollary 1.6.2

Let $\hat{V}_{uo}^{(p)}$ be an unordered unbiased estimator of $V\left(\hat{T}_{uo}^{(p)}\right)$. Then, an unbiased estimator $\hat{V}_{cu}^{(l)}$ of the variance of $\hat{T}_{cu}^{(l)}$ is given by

$$\hat{V}_{cu}^{(l)} = \left(\hat{T}_{cu}^{(l)}\right)^2 - \sum_{p=1}^{M_{kl}} \left\{\left(\hat{T}_{uo}^{(p)}\right)^2 - \hat{V}_{uo}^{(p)}\right\} P'(S_p) \tag{1.6.7}$$

Proof :

$$E\left(\hat{V}_{cu}^{(l)}\right) = E\left(\hat{V}_{cu}^{(l)}\right)^2 - \sum_{p=1}^{M_{kl}} E\left(\hat{T}_{uo}^{(p)}\right)^2 P'(S_p) + \sum_{p=1}^{M_{kl}} E\left(\hat{V}_{uo}^{(p)}\right) P'(S_p)$$

$$= E\left(\hat{T}_{cu}^{(l)}\right)^2 - E\left(\hat{T}_{uo}^{(p)}\right)^2 + E\left(\hat{V}_{uo}^{(p)}\right)$$

$$= E\left(\hat{T}_{cu}^{(l)}\right)^2 - E\left(\hat{T}_{uo}^{(p)}\right)^2 + V\left(\hat{T}_{uo}^{(p)}\right)$$

$$= E\left(\hat{T}_{cu}^{(l)}\right)^2 - \left\{E\left(\hat{T}_{uo}^{(p)}\right)\right\}^2$$

$$= E\left(\hat{T}_{cu}^{(l)}\right)^2 - T^2$$

$$= V\left(\hat{T}_{cu}^{(l)}\right)$$

1.7 APPLICATIONS OF UNORDERING AND COMBINED UNORDERING

It is shown in the forgoing discussion that the above techniques give, in general, improved estimators, in the sense of efficiency, when applied to original estimators. We know that T_1, T_3, T_4, T_6 and T_7-class estimators, in general, are ordered estimators and so they can be unordered. Further T_2-class estimators for sampling with replacement, though unordered, but can be combined unordered. In the following section, we discuss unordering of T_7-class for illustration purpose. After this we discuss in Section 1.9 combined unordering of the classical SRSWR estimator, which is unordered to start with.

1.8 UNORDERING OF T_7-CLASS OF LINEAR ESTIMATOR

We know that an ordered T_7-class of estimator is given by $\hat{T}_7 = \sum \beta_{ro}^{s_t} X_r$, where in general t = 1, 2, . . ., $\sum_{k=1}^{n} \sum_{l=1}^{\binom{N}{k}} \sum_{p=1}^{M_{kl}} M_{klp}$.

The unordered estimator is

$$\left(\hat{T}_{uo}\right)=\frac{\sum\limits_{t=1}^{M_{klp}} p(s_t)\left(\sum\limits_{r=1}^{n} \beta_{ro}^{s_t} X_r\right)}{\sum\limits_{t=1}^{M_{klp}} p(s_t)} \tag{1.8.1}$$

The numerator of equation (1.8.1) is

$$= \sum_{t=1}^{M_{klp}} p(s_t)\left(\sum_{r=1}^{n} \beta_{ro}^{s_t} X_r\right)$$

$$= \sum_{j=1}^{k} x_{i_j} \left\{\sum_{t=1}^{M_{klp}} p(s_t) \sum_{r=1}^{n} \beta_{r,i_j}^{s_t} \gamma_r^{s_t} (i_j)\right\}$$

where,

$$\gamma_r^{s_t}(i_j) = \begin{cases} 1, \text{ if } i_j-\text{th unit occurs at the } r-\text{th draw in the } s_t-\text{th sample} \\ 0, \text{ otherwise} \end{cases}$$

$$\text{Let } \beta_{i_j}^{p} = \left\{\sum_{t=1}^{M_{klp}} p(s_t) \sum_{r=1}^{n} \beta_{r,i_j}^{s_t} \gamma_r^{s_t}(i_j)\right\} \frac{1}{\sum\limits_{t=1}^{M_{klp}} p(s_t)}$$

$$\left(\hat{T}_{uo}^{(p)}\right) = \sum_{i_j \in s_t} x_{i_j} \beta_{i_j}^{p} \tag{1.8.2}$$

Thus we observe that the estimator given in Eq. (1.8.2) belongs to T_5-class. That, this estimator can be further combined unordered in the same manner, is clear.

1.9 COMBINED UNORDERING OF THE CLASSICAL SRSWR ESTIMATOR

For simple random sampling with replacement, the classical estimator of the population total, which is unordered, can be written as

$$\hat{T}_{uo}^{(p)} = \frac{N}{n}\left(\sum_{j=1}^{k} x_{i_j} R_{i_j}^{(p)}\right) \tag{1.9.1}$$

Let $P'(S_p) = P(S_p) / \sum\limits_{p=1}^{M_{kl}} P'(S_p)$. Then by combining such M_{kl} unordered estimators, we have from Eq. (1.6.1), the combined unordered estimator as

$$\hat{T}_{cu}^{(l)} = \sum_{p=1}^{M_{kl}} \left(\hat{T}_{uo}^{(p)}\right)_{S_p} P'(S_p)$$

$$= \sum_{p=1}^{M_{kl}} \frac{N}{n} \left\{\left(\sum_{j=1}^{k} x_{i_j} R_{i_j}^{(p)}\right) P'(S_p)\right\}$$

$$= \frac{N}{n} \sum_{j=1}^{k} x_{i_j} \left\{\sum_{p=1}^{M_{kl}} R_{i_j}^{(p)} . P'(S_p)\right\}$$

$$= \frac{N}{n} \sum_{j=1}^{k} x_{i_j} \left\{\sum_{p=1}^{M_{kl}} R_{i_j}^{(p)} . P(S_p) / \left(\sum_{p=1}^{M_{kl}} P(S_p)\right)\right\}$$

Let R_{i_j} denote the number of times the i_j-th unit occurs in a sample. Then

$$E\left(R_{i_j} | \zeta_l\right) = \frac{\sum_{p=1}^{M_{kl}} R_{i_j}^{(p)} . P(S_p)}{\sum_{p=1}^{M_{kl}} P(S_p)}$$

Also,

$$E\left(R_{i_j} | \zeta_l\right) = E\left\{\sum_{r=1}^{n} \gamma_{i_j}(r) | \zeta_l\right\}$$

where,

$$\gamma_{i_j}(r) = \begin{cases} 1, & \text{if } i_j - \text{th unit occurs at the } r - \text{th draw in the sample} \\ 0, & \text{otherwise} \end{cases}$$

Now,

$$E\left\{\sum_{r=1}^{n} \gamma_{i_j}(r) | \zeta_l\right\} = \sum_{r=1}^{n} E\left\{\gamma_{i_j}(r) | \zeta_l\right\}$$

$$= \sum_{r=1}^{n} \frac{1}{k} = \frac{n}{k}$$

So,

$$E\left(R_{i_j} | \zeta_l\right) = \frac{n}{k}.$$

Thus

$$\hat{T}_{cu}^{(l)} = \frac{N}{n}\left(\sum_{j=1}^{k} x_{i_j}\right)\frac{n}{k}$$

$$= N.\,\overline{x}_k$$

So, $\overline{x}_k = \left(\sum_{j=1}^{k} x_{i_j}\right)/k$ is an improved estimator over $\overline{x}_n = \left(\sum_{j=1}^{k} x_{i_j} R_{i_j}\right)/n$ for estimating the population mean. This estimator was obtained earlier by Basu (1958) by a difficult method using the concept of sufficiency in survey sampling.

It may be noted, by taking $\beta_{i_j}^{S_p} = 1/(k.R_{i_j})$ that the above estimator belongs to T_5-class. Since it is shown in Chapter 6 that MVLUE does not exist in T_5-class for any sampling scheme, the above estimator is not MVLUE in this class.

Before we proceed to obtain the variance of the above estimator, we establish the following Lemma giving probability of occurrence of a class ζ_l.

Lemma 1.9.1

The probability of occurrence of a class ζ_l, when there is simple random sampling with replacement, is given by

$$P(\zeta_l) = \frac{1}{N^n}\left\{\sum_{\nu=0}^{k-1}(-1)^{\nu}\binom{k}{\nu}(k-\nu)^n\right\}$$

Proof :

The occurrence of a sample in which i_j-th unit occurs R_{i_j} times for $j = 1, 2, \ldots, k$, such that $\sum_{j=1}^{k} R_{i_j}^{(p)} = n$ is equivalent to occurrence first of an event in which k distinct units are selected and then of the event in which k cells, representing the selected k distinct units are filled up by n balls representing n draws in such a manner that no cell remains empty. Now, the probability of the event in which some cells are empty is given by

$$p(\nu) = \frac{(k - \nu)^n}{k^n} \quad \text{for } \nu = 1, 2, \ldots, k.$$

Let,
$$S_\nu = \binom{k}{\nu} (k - \nu)^n \frac{1}{k^n}$$

Let P_1 denote the probability of the event in which atleast one cell is empty. The P_1 is given by (Feller, 1972, p. 99, III edition, Wiley Eastern Private Limited, New Delhi).

$$P_1 = \sum_{\nu=1}^{k} \binom{k}{\nu} (k - \nu)^n \frac{1}{k^n} (-1)^{\nu-1}$$

Therefore, the required probability of the event in which no cell is empty is given by

$$1 - P_1 = 1 - \sum_{\nu=1}^{k} \binom{k}{\nu} (k - \nu)^n \frac{1}{k^n} (-1)^{\nu-1}$$

$$= \sum_{\nu=0}^{k-1} \binom{k}{\nu} (k - \nu)^n \frac{1}{k^n} (-1)^{\nu}$$

Thus,
$$P(\zeta_l) = (1 - P_1) \cdot \frac{k^n}{N^n},$$

as the probability of the event, in which k distinct units are selected, is k^n/N^n. Therefore,

$$P(\zeta_l) = \frac{1}{N^n} \left\{ \sum_{\nu=0}^{k} (-1)^\nu \binom{k}{\nu} (k - \nu)^n \right\}$$

Now we proceed to obtain the variance of the estimator $\hat{T}_{cu}^{(l)}$. Using Eq. (1.6.4), we have

$$V\left(\hat{T}_{cu}^{(l)}\right) = \sum_{k=1}^{n} \sum_{l=1}^{\binom{N}{k}} \left\{ \frac{N}{k} \left(\sum_{j=1}^{k} x_{i_j} \right) \right\}^2 P(\zeta_l) - T^2$$

$$= N^2 \sum_{i=1}^{N} x_i^2 \left\{ \sum_{k=1}^{n} \frac{1}{k^2} \Sigma' P(\zeta_l) \right\}$$

$$+ 2N^2 \cdot \sum_{i>j} x_i x_j \left\{ \sum_{k=1}^{n} \frac{1}{k^2} \Sigma'' P(\zeta_l) \right\} - T^2 \tag{1.9.3}$$

where, $\sum'$ denotes sum over $\binom{N-1}{k-1} - \zeta_l$ classes in which a given i-th unit occurs and $\sum''$ denotes the sum over $\binom{N-2}{k-2} - \zeta_l$ classes in which a pair (i, j) occurs.

We note from Lemma 1.9.1 that $P(\zeta_l)$ depends only on k and n and not on the nature of units. Therefore, Eq. (1.9.3) gives

$$V\left(\hat{T}_{cu}^{(l)}\right) = \left[N^2 . \left\{\sum_{i=1}^{N} x_i^2 \left(\sum_{k=1}^{n} \frac{1}{k^2} . \binom{N-1}{k-1} . P(\zeta_l)\right)\right\}\right.$$

$$\left. + 2N^2 . \left\{\sum_{i<j}^{N} x_i x_j \left(\sum_{k=1}^{n} \frac{1}{k^2} . \binom{N-2}{k-2} P(\zeta_l)\right)\right\}\right] \qquad (1.9.4)$$

Pathak (1965, Eq. 16, p. 16) has also given the variance of the above estimator in the following form

$$V\left(\hat{T}_{cu}\right)_{pathak} = \left[N^2 \left\{1^{n-1} + 2^{n-1} + ... + (N-1)^{n-1}\right\} . S^2\right] / N^n \qquad (1.9.5)$$

where, $S^2 = \sum_{i=1}^{N} (x_i - \overline{x}_N)^2 / (N-1)$. From comparison of the two forms in special cases, it appears that these forms of the variance would give the same value.

EXERCISES

1.1 Discuss Partition of sample space given by Tikkiwal (1972) and illustrate it by taking an example. Hence, define the modified set of T-classes of linear estimators given by Bhargava and Tikkiwal (1973).

1.2 Discuss T_5-class of linear estimators given by Bhargava and Tikkiwal (1973). Further, show that an estimator in any one of the following classes lies in the T_5-class :

(i) $T_{2(GK)}$-class due to Godambe and Koop.

(ii) $T_{3(K)}$- class due to Koop.

1.3 Let a sample of size 2 is selected with varying probabilities without replacement scheme from a finite population of size N. Further, let (x_1, x_2) and (p_1, p_2) are respectively the x-values of the sample units, selected at the first and second draw, and their initial probabilities of selection.

If $\quad t_1 = \frac{x_1}{p_1}, t_2 = x_1 + \frac{x_2}{p_2}(1-p_1)$

and $\quad t = \frac{t_1 + t_2}{2}$, then show that the ordered estimator t is an unbiased estimator of population total T, and

$$V(t) = \left(1 - \frac{1}{2}\sum_{i=1}^{N} p_i^2\right)\left\{\frac{1}{2}\sum_{i=1}^{N} p_i\left(\frac{x_i}{p_i} - T\right)^2\right\} - \frac{1}{4}\sum_{i=1}^{N} p_i^2\left(\frac{x_i}{p_i} - T\right)^2 \tag{1}$$

Further, show that an unbiased estimator of V(t) is

$$v(t) = \frac{(1-p_1)^2}{4}\left(\frac{x_1}{p_1} - \frac{x_2}{p_2}\right)^2 \tag{2}$$

[Des Raj (1956)]

1.4 Let $(u_1, u_2, \ldots, u_n)$ be the units selected with varying proabilities without replacement and in the order in which they are drawn from a finite population of size N. Further let $(x_1, \ldots, x_n)$ and $(p_1, p_2, \ldots, p_n)$ are respectively the x-values of the sample units and their initial probabilities of selection. Define $t_1 = x_1/p_1$,

$$t_i = x_1 + x_2 + \ldots + x_{i-1} + \frac{x_i}{p_i}(1 - p_1 - p_2 - \ldots - p_{i-1}), \qquad i = 2, \ldots, n$$

and $t = \frac{1}{n}\sum_{i=1}^{n} t_i$

Then show that :

(a) estimators t_i $(i = 1, \ldots, n)$ are each unbiased of population total T and are mutually uncorrelated.

(b) $V(t) \le V(\hat{T}_{pps})$, where $\hat{T}_{pps} = \sum_{i=1}^{n} x_i/np_i$ the Hansen-Hurwitz's estimator of population total under PPSWR scheme.

(c) an unbiased estimator of $V(t)$ is $v(t) = \frac{1}{n(n-1)}\sum_{i=1}^{n}(t_i - t)^2$

[Des Raj (1956)]

1.5 Explain Murthy's (1957) technique of unordering an ordered estimator when there is sampling with varying probabilities without replacement. Further, show that Murthy's unordered estimator is more efficient than the corresponding ordered estiamtor.

1.6 Show that for $n=2$ the Murthy's unordered estimator corresponding to Des Raj's ordered estimator t, defined in Exercise 1.3 is given by

$$\hat{T}_M = \frac{1}{2-p_1-p_2}\left[(1-p_2)\frac{x_1}{p_1}+(1-p_1)\frac{x_2}{p_2}\right]$$

The variance of this estimator is

$$V\left(\hat{T}_M\right)=\sum_{i=1}^{N}\frac{x_i^2}{p_i(1-p_i)}\sum_{j(\neq i)=1}^{N}\frac{p_j(1-p_j)}{2-p_i-p_j}+\sum_{i=1}^{N}\sum_{j(\neq i)=1}^{N}\frac{x_i x_j}{2-p_i-p_j}-T^2$$

and an unbiased estimator of $V\left(\hat{T}_M\right)$ is

$$\hat{V}\left(\hat{T}_M\right)=\frac{(1-p_1)(1-p_2)(1-p_1-p_2)}{(2-p_1-p_2)^2}\left(\frac{x_1}{p_1}-\frac{x_2}{p_2}\right)^2$$

1.7 Discuss unified approach for unordering an ordered estimator for sampling with varying proabilities with or without replacement, given by Tikkiwal (1972). If $\hat{T}_{uo}^{(p)}$ denotes the unordered estimator corresponding to ordered estimator $\hat{T}_{s_t}$ of population total T, then show that :

(a) $E\left(\hat{T}_{uo}^{(p)}\right) = E\left(\hat{T}_{s_t}\right)$

(b) $V\left(\hat{T}_{uo}^{(p)}\right) \leq V\left(\hat{T}_{s_t}\right)$

Further, show that Murthy's (1957) results for sampling with varying proabilities without replacement can be obtained as a special case.

1.8 Discuss Tikkiwal's (1972) technique of combined unordering for sampling with varying probabilities with replacement. Further, show that:

$$E\left(\hat{T}_{cu}^{(l)}\right) = E\left(\hat{T}_{uo}^{(p)}\right) \qquad (1)$$

and $$V\left(\hat{T}_{cu}^{(l)}\right) \leq V\left(\hat{T}_{uo}^{(p)}\right) \qquad (2)$$

where, $\hat{T}_{cu}^{(l)}$ is a combined unordered estimator based on M_{kl} unordered estimators $\hat{T}_{uo}$

1.9 Show that the combined unordering of the classical SRSWR estimator $\overline{X}_n$ (the sample mean) gives the estimator $\overline{X}_k$ (the mean of k-distinct units in the sample). Further, show that:

(a) $\overline{X}_k$ is an unbiased estimator of population mean $\overline{x}_N$,

(b) $V(\overline{X}_k) = \dfrac{S^2}{N} \sum\limits_{j=1}^{N-1} \left(\dfrac{j}{N}\right)^{n-1}$, and

(c) $V(\overline{X}_k) \le V(\overline{X}_n)$

where, S^2 is population mean square.

[Tikkiwal, G.C. and Jain, V. (2001); Pathak (1965)]

1.10 Show that the two alternative unbiased estimators of $V(\overline{X}_k)$ given in Example 1.9 are :

(a) $v_1(\overline{X}_k) = \left[\left(\dfrac{1}{k} - \dfrac{1}{N}\right) + N^{1-n}\left(1 - \dfrac{1}{k}\right)\right] s_k^2$ for $k \ge 2$

where, $s_k^2 = \dfrac{1}{k-1} \sum\limits_{i \varepsilon s} (x_i - \overline{X}_k)^2$

(b) $v_2(\overline{X}_k) = \left[\left(\dfrac{1}{k} - \dfrac{1}{N}\right) + \dfrac{N-1}{N^n - N}\right] s_o^2$

where, $s_o^2 = \begin{cases} s_k^2 & \text{for } k \ge 2 \\ 0 & \text{for } k = 1 \end{cases}$

[Des Raj and Khamis (1958)]

CHAPTER 2

T_1 - Class Estimators

2.1 INTRODUCTION

Following the convention of Section 1.1 for the study of a variate let there be a population $\pi:(x_1, x_2, \ldots, x_N)$ of size N units. Let us draw from it a random sample of size n. The random sample, itself, is drawn with the help of some sampling schemes.

In this chapter, we present results for T_1-class when the random sample is drawn by any one of the SRSWOR, SRSWR, Midzuno and Horvitz-Thompson sampling schemes, and the general varying probability sampling scheme with replacement. These results are due to Prabhu Ajgaonkar and Tikkiwal (1961) originally given in abstract form, subsequently in detail by former author (1962, section 1.3 and 1.4, pp. 12-24, section 1.8, pp. 48-51 and 1967). Because of the importance of the results the abstract is given in Appendix I for ready reference.

From these results, it will be seen that varying probability sampling schemes, as against simple random sampling schemes, do not always lead to more efficient estimators, as generally believed.

In particular it is noted that Midzuno sampling scheme with replacement, a varying probability scheme yields poorer unbiased estimators in this class than SRSWR scheme does. Further it is shown that there does not exist any unbiased estimator in this class for the two varying probability sampling schemes: (i) Horvitz - Thompson sampling scheme and (ii) The general varying probability sampling scheme with replacement.

2.2 ESTIMATORS IN T_1-CLASS FOR SOME WELL KNOWN SAMPLING SCHEMES

We first present a Lemma which is useful for obtaining results in this class. Let p_{ir} be the probability of drawing i-th unit at r-th draw for i= 1, 2, . . ., N and r = 1, 2, . . . , n for a specified sampling scheme. Let $p(j,s \mid i, r)$ be the conditional probability of drawing j-th unit at the s-th draw given that i-th unit has occured at the r-th draw. Let a_r denote the weight to be associated with X_r the random variable for r-th draw. Let the probability matrix

$$P = \begin{bmatrix} p_{11} & p_{12} & \cdots & p_{1n} \\ \cdot & & & \\ \cdot & & & \\ p_{N1} & p_{N2} & \cdots & p_{Nn} \end{bmatrix}$$

for the specified probability system of the sampling procedure.

Let

$$P' = \begin{bmatrix} p_{11} & p_{12} & \cdots & p_{1n} & 1 \\ p_{21} & p_{22} & \cdots & p_{2n} & 1 \\ \cdot & & & & \\ \cdot & & & & \\ \cdot & & & & \\ p_{N1} & \cdots & \cdots & p_{Nn} & 1 \end{bmatrix}$$

denote another matrix then we prove the following Lemma for the existence of T_1 - class of linear estimators as defined in Section 1.2 of Chapter 1.

Lemma 2.2.1

There exists unbiased estimators, of the population total T, of the form $\hat{T} = \sum_{r=1}^{n} a_r x_r$ of the population total T, for any specified sampling scheme, if the rank of the matrix P is the same as that of the matrix P'. The variance of $\hat{T}$ is given by

$$V(\hat{T}) = \sum_{r=1}^{n} a_r^2 \, V(X_r) + 2 \sum_{r=1}^{n} \sum_{s>r}^{n} a_r \, a_s \, Cov(X_r, X_s)$$

where,

$$V(X_r) = \sum_{i=1}^{N} x_i^2 \, p_{ir} - \left(\sum_{i=1}^{N} x_i \, p_{ir} \right)^2$$

and,

$$Cov(X_r, X_s) = \sum_{i=1}^{N} \sum_{j=1}^{N} x_i \, x_j \, p_{ir} \, p(j, s \mid i, r) - \left(\sum_{i=1}^{N} x_i \, p_{ir} \right) \left(\sum_{j=1}^{N} x_j \, p_{js} \right)$$

Proof :

Let our estimator for the population total be

$$\hat{T} = \sum_{r=1}^{n} a_r X_r$$

Then,

$$E\left(\hat{T}_1\right) = E\left(\sum_{r=1}^{n} a_r X_r\right)$$

$$= \sum_{r=1}^{n} a_r E(X_r)$$

$$= \sum_{r=1}^{n} a_r \sum_{i=1}^{N} x_i p_{ir}$$

$$= \sum_{i=1}^{N} x_i \left(\sum_{r=1}^{n} a_r p_{ir}\right) \qquad (2.2.1)$$

For the unbiasedness of $\hat{T}_1$ for any arbitrary population total T, it is necessary and sufficient that

$$\sum_{r=1}^{n} a_r p_{ir} = 1 \qquad \text{for } i = 1, 2, \ldots, N \qquad (2.2.2)$$

The Eq. (2.2.2) can be written in the matrix form as

$$P.A = I \qquad (2.2.3)$$

where,

$$P = \begin{bmatrix} p_{11} & \cdots & p_{1n} \\ p_{21} & \cdots & p_{2n} \\ . & & \\ . & & \\ p_{N1} & \cdots & p_{Nn} \end{bmatrix}_{N\text{x}n} ; \quad A = \begin{bmatrix} a_1 \\ a_2 \\ . \\ . \\ a_n \end{bmatrix}_{n\text{x}1} \quad \text{and } I = \begin{bmatrix} 1 \\ 1 \\ . \\ . \\ 1 \end{bmatrix}_{N\text{x}1}$$

In order that the equations in the system (2.2.3) are consistent, it is necessary as well as sufficient to have

$$R(P) = R(P')$$

otherwise the `Non-Homogenious Linear Equations' in the system (2.2.3) are unsolvable. This proves the first part of the Theorem. Again,

$$V\left(\hat{T}\right) = \text{Var}\left(\sum_{r=1}^{n} a_r X_r\right)$$

$$= \sum_{r=1}^{n} \text{Var}(a_r X_r) + 2 \sum_{r=1}^{n} \sum_{s>r}^{n} \text{Cov}(a_r X_r, a_s X_s)$$

$$= \sum_{r=1}^{n} a_r^2 \text{Var}(X_r) + 2 \sum_{r=1}^{n} \sum_{s>r}^{n} a_r a_s \text{Cov}(X_r, X_s) \qquad (2.2.4)$$

where,

$$\text{Var}(X_r) = E(X_r)^2 - \{E(X_r)\}^2$$

$$= \sum_{i=1}^{N} x_i^2 p_{ir} - \left(\sum_{i=1}^{N} x_i p_{ir}\right)^2 \qquad (2.2.5)$$

$$\text{Cov}(X_r, X_s) = E(X_r X_s) - E(X_r) E(X_s)$$

$$= \sum_{i=1}^{N} \sum_{j=1}^{N} x_i x_j p_{ir} p(j,s \mid i, r) - \left(\sum_{i=1}^{N} x_i p_{ir}\right)\left(\sum_{j=1}^{N} x_j p_{js}\right) \qquad (2.2.6)$$

Hence the Lemma.

(a) Simple Random Sampling Schemes : SRSWOR and SRSWR

We proceed to study the existence of MVLUE first in case of simple random sampling without replacement (SRSWOR). The main results are given in the following theorem. These results are also presented in Sukhatme and Sukhatme (1970, Ex. 1.6, p. 39).

Theorem 2.2.1

The MVLUE exists in T_1-class for SRSWOR is given by $\hat{T} = \frac{N}{n} \sum_{r=1}^{n} X_r$. The variance of the estimator is given by $\frac{N(N-n)}{n} S^2$, where $S^2 = \sum_{i=1}^{N} (x_i - \bar{x}_N)^2/(N-1)$.

Proof :

From Lemma 1.3.1 for SRSWOR,

$$p_{ir} = \frac{1}{N}$$

for $i = 1, 2, \ldots, N$ and $r = 1, 2, \ldots, n$.

Therefore, the rank of the matrix

$$P = \begin{bmatrix} \frac{1}{N} & \cdots & \frac{1}{N} \\ . & & \\ . & & \\ . & & \\ \frac{1}{N} & \cdots & \frac{1}{N} \end{bmatrix} \tag{2.2.7}$$

is one which is also the rank of the matrix

$$P' = \begin{bmatrix} \frac{1}{N} & \cdots & \frac{1}{N} & 1 \\ \frac{1}{N} & \cdots & \frac{1}{N} & 1 \\ . & & & \\ . & & & \\ \frac{1}{N} & \cdots & \frac{1}{N} & 1 \end{bmatrix} \tag{2.2.8}$$

Thus by Lemma 2.2.1, there exists unbiased estimators in T_1-class.

The effective number of conditions for $\hat{T}$ to be unbiased estimator for population total T is equal to the rank of P. Since the rank of P is one, we have this effective condition as

$$\sum_{r=1}^{n} a_r = N \tag{2.2.9}$$

$$\text{Var}(X_r) = \sum_{i=1}^{N} x_i^2 \, p_{ir} - \left(\sum_{i=1}^{N} x_i \, p_{ir} \right)^2$$

$$= \frac{1}{N} \sum_{i=1}^{N} x_i^2 - \bar{x}_N^2$$

$$= \frac{1}{N} \sum_{i=1}^{N} (x_i - \bar{x}_N)^2$$

$$= \frac{1}{N} (N-1)\, S^2, \text{ where } S^2 = \sum_{i=1}^{N} (x_i - \bar{x}_N)^2/(N-1) \tag{2.2.10}$$

$$\text{Cov}(X_r, X_s) = \sum_{i=1}^{N} \sum_{j=1}^{N} x_i x_j \, p_{ir} \, p(j, s \mid i, r) - \left(\sum_{i} x_i \, p_{ir} \right) \left(\sum_{j=1}^{N} x_j \, p_{js} \right)$$

Since,

$$p(j,s|i,r) = \begin{cases} \dfrac{1}{N-1}, & \text{for } j \neq i \\ 0 \quad , & \text{for } j = i \end{cases}$$

for SRSWOR scheme

$$\text{Cov}\,(X_r, X_s) = \sum_{i=1}^{N} \sum_{j(\neq i)}^{N} x_i x_j \frac{1}{N}\frac{1}{N-1} - \bar{x}_N^2$$

$$= \frac{1}{N(N-1)} \left\{ \left(\sum_{i=1}^{N} x_i \right)^2 - \sum_{i=1}^{N} x_i^2 \right\} - \bar{x}_N^2$$

$$= \frac{1}{N(N-1)} \left(N\,\bar{x}_N^2 - \Sigma x_i^2 \right)$$

$$= -\frac{1}{N(N-1)} \sum_{i=1}^{N} (x_i - \bar{x}_N)^2$$

$$= -\frac{1}{N} S^2 \tag{2.2.11}$$

Therefore, by Lemma 2.2.1

$$\text{Var}\left(\hat{T}_1\right) = \sum_{r=1}^{n} a_r^2 \left(\frac{N-1}{N} \right) S^2 - \frac{2S^2}{N} \sum_{r=1}^{n} \sum_{s>r}^{n} a_r a_s$$

$$= \left(\frac{N-1}{N} \right) S^2 \left(\sum_{r=1}^{n} a_r^2 \right) - \frac{S^2}{N} \sum_{r=1}^{n} \sum_{s(\neq r)=1}^{n} a_r a_s$$

$$= \left(\frac{N-1}{N} \right) S^2 \left(\Sigma\, a_r^2 \right) - \frac{S^2}{N} \left\{ (\Sigma a_r)^2 - \Sigma\, a_r^2 \right\}$$

$$= S^2 \left(\Sigma\, a_r^2 \right) - \frac{S^2}{N} (\Sigma a_r)^2 \tag{2.2.12}$$

Thus, to find the condition that $\hat{T}_1$ is the minimum variance linear unbiased estimator (MVLUE) of T, we minimise the variance of $\hat{T}_1$ with respect to the condition given by Eq. (2.2.9). For this we construct the function.

$$\phi = S^2 \left(\Sigma\, a_r^2 \right) - \frac{S^2}{N} (\Sigma\, a_r)^2 - 2\lambda\,(\Sigma\, a_r)$$

where, λ is Lagrange's multiplier. Differentiating ϕ with respect to a_r and equating it to zero we get

$$2S^2\ a_r - \frac{2S^2}{N}(\Sigma\ a_r) - 2\lambda = 0 \tag{2.2.13}$$

Using Eq. (2.2.9), we get

$$\lambda = (a_r - 1)\ S^2$$

Summing over r = 1, 2, . . ., n we get

$$\lambda = \frac{N-n}{n}\ S^2 \tag{2.2.14}$$

For $S^2 \neq 0$, which gives

$$a_r = \frac{N}{n} \tag{2.2.15}$$

For $S^2 = 0$, $x_i = \overline{x}_N$ for $i = 1, 2, . . ., N$ and so inference about mean is easily made by drawing a sample of size one and so we do not discuss it further. Thus, when $S^2 \neq 0$ the condition for $\hat{T}_1$ to be the minimum variance linear unbiased estiator for population total is that the weight associated to the outcome of each draw is the constant $= N/n$. Hence, $\hat{T}_1$ is given by

$$\hat{T}_1 = \frac{N}{n} \sum_{r=1}^{n} X_r \tag{2.2.16}$$

From Eqs. (2.1.10) and (2.2.15)

$$V\left(\hat{T}_1\right) = \frac{N^2}{N} \frac{(N-n)}{n}\ S^2$$

Hence, the Theorem.

For SRSWR, from Lemma 2.2.1 we note that P matrix is the same as in Eq. (2.2.7) and therefore, the Eq. (2.2.9) holds in this case also. Further, $V(X_r)$ remains the same as in Eq. (2.2.10) and $Cov\ (X_r, X_s) = 0$ and so $V\left(\hat{T}_1\right)$ is obtained from Lemma 2.2.1. From these results, we then easily obtain the following theorem.

Theorem 2.2.2

The MVLUE exists in T_1-class for SRSWR and is given by $\hat{T}_1 = \frac{N}{n} \sum_{r=1}^{n} X_r$, the variance of the estimator is $\{N^2 (N-1)/Nn\} S^2$.

(b) Midzuno Scheme of Sampling With or Without Replacement

We present, first, the results for sampling without replacement in the following Theorem.

Theorem 2.2.3

The MVLUE exists in T_1-class for Midzuno Scheme of Sampling without replacement and is given by

$$\hat{T}_1 = X_1 + \frac{N-1}{n-1} \sum_{r=2}^{n} X_r$$

Its variance is given by

$$V\left(\hat{T}_1\right) = \frac{N-n}{n-1}\left\{\sum_{i=1}^{N} (1-p_i)\, x_i^2 - \frac{1}{N-2} \sum_{i \neq j} (1-p_i-p_j)\, x_i x_j\right\}$$

Proof :

Let us, derive the estimator of population total and its variance in Midzuno sampling scheme first without replacement.

For this scheme we have $p_{i1} = p_i$ for all i

$$\begin{aligned} p_{ir} &= p\,(i' \neq i, 1)\; p\left\{(i' \neq i,2) \mid (i' \neq i, 1)\right\} \ldots p\left\{(i' \neq i, r-1) \mid (i' \neq i, 1)\right. \\ &\quad \left.(i' \neq i, 2), \ldots, (i' \neq i, r-2)\right\}\; p\left\{(i,r) \mid (i' \neq i,1), \ldots, (i' \neq i, r-1)\right\} \\ &= (1-p_i)\left(1-\frac{1}{N-1}\right) \cdots \left(1-\frac{1}{N-r+2}\right) \cdot \frac{1}{N-r+1} \\ &= (1-p_i)\,\frac{N-2}{N-1}\,\frac{N-3}{N-2} \cdots \frac{N-r+1}{N-r+2} \cdot \frac{1}{(N-r+1)} \\ &= \frac{(1-p_i)}{(N-1)} \qquad \text{for all i and } r = 2, \ldots, n. \end{aligned} \tag{2.2.17}$$

Thus the probability matrix for the Midzuno sampling scheme without replacement is given by

$$P = \begin{bmatrix} p_1 & \frac{1-p_1}{N-1} & \cdots & \frac{1-p_1}{N-1} \\ p_2 & \frac{1-p_2}{N-1} & \cdots & \frac{1-p_2}{N-1} \\ . & & & \\ . & & & \\ . & & & \\ p_N & \frac{1-p_N}{N-1} & \cdots & \frac{1-p_N}{N-1} \end{bmatrix} \tag{2.2.18}$$

The condition for existence of an unbiased Midzuno's Estimator for population total is that the rank of this matrix is same as that of P', where

$$P' = \begin{bmatrix} p_1 & \frac{1-p_1}{N-1} & \cdots & \frac{1-p_1}{N-1} & 1 \\ p_2 & \frac{1-p_2}{N-1} & \cdots & \frac{1-p_2}{N-1} & 1 \\ . & & & & \\ . & & & & \\ . & & & & \\ p_N & \frac{1-p_N}{N-1} & \cdots & \frac{1-p_N}{N-1} & 1 \end{bmatrix} \tag{2.2.19}$$

Subtracting second column from third, fourth, fifth and onward, we get

$$R(P) = R\begin{bmatrix} p_1 & \frac{1-p_1}{N-1} & 0 & 0 \ldots 0 \\ p_2 & \frac{1-p_2}{N-1} & 0 & \ldots 0 \\ . & & & \\ . & & & \\ . & & & \\ p_N & \frac{1-p_N}{N-1} & 0 & \ldots 0 \end{bmatrix} \tag{2.2.20}$$

All the p_i's are not equal as otherwise the above scheme will become SRSWOR. Thus without any loss of generality we can assume $p_1 \neq p_2$ and therefore at least one of them is not zero and let it be p_1. We consider the following leading principal minor of order 2 from matrix Eq. (2.2.20) :

$$\Delta_{12} = \begin{vmatrix} p_1 & \dfrac{1-p_1}{N-1} \\ p_2 & \dfrac{1-p_2}{N-1} \end{vmatrix} = \frac{1}{N-1}\left\{(p_1 - p_1 p_2) - (p_2 - p_1 p_2)\right\}$$

$$= \frac{1}{N-1}(p_1 - p_2) \neq 0$$

Thus R (P) = 2. We also have

$$R(P') = R\begin{bmatrix} p_1 & \dfrac{1-p_1}{N-1} & \cdots & \dfrac{1-p_1}{N-1} & 1 \\ p_2 & \dfrac{1-p_2}{N-1} & \cdots & \dfrac{1-p_2}{N-1} & 1 \\ \cdot & \cdot & & \cdot & \cdot \\ \cdot & \cdot & & \cdot & \cdot \\ \cdot & \cdot & & \cdot & \cdot \\ p_N & \dfrac{1-p_N}{N-1} & \cdots & \dfrac{1-p_N}{N-1} & 1 \end{bmatrix}$$

Add (N – 1) multiple of column second to the first and subtracting the resultant from last column we get

$$R(P') = R\begin{bmatrix} p_1 & \dfrac{1-p_1}{N-1} & \cdots & \dfrac{1-p_1}{N-1} & 0 \\ p_2 & \dfrac{1-p_2}{N-1} & \cdots & \dfrac{1-p_2}{N-1} & 0 \\ \cdot & \cdot & & \cdot & \cdot \\ \cdot & \cdot & & \cdot & \cdot \\ \cdot & \cdot & & \cdot & \cdot \\ p_N & \dfrac{1-p_N}{N-1} & \cdots & \dfrac{1-p_N}{N-1} & 0 \end{bmatrix}$$

Again subtracting second column from n^{th}, $(n\text{-}1)^{th}$, . . . third we get

$$R(P') = R\begin{bmatrix} p_1 & \dfrac{1-p_1}{N-1} & 0 \ldots 0 \\ p_2 & \dfrac{1-p_2}{N-1} & 0 \ldots 0 \\ \cdot & & \\ \cdot & & \\ \cdot & & \\ p_N & \dfrac{1-p_N}{N-1} & 0 \ldots 0 \end{bmatrix}$$

Here also

$$\Delta_{12} = \begin{vmatrix} p_1 & \dfrac{1-p_1}{N-1} \\ p_2 & \dfrac{1-p_2}{N-1} \end{vmatrix} \neq 0$$

Hence R (P') = 2.

The condition of unbiasedness is as in matrix Eq. (2.2.3). Since R (P) = R (P') = 2, we have only two effective equations. To get these equations, let us apply the following transformation where multiplying the first row by p_i/p_1 (i = 2, . . ., N) and then subtracting it from the i-th row ($i \geq 2$) in (2.2.3) with P as in Eq. (2.2.18). Then,

$$\begin{bmatrix} p_1 & \dfrac{1-p_1}{N-1} & \cdots & \dfrac{1-p_1}{N-1} \\ 0 & S_2 & \cdots & S_2 \\ \cdot & & & \\ \cdot & & & \\ \cdot & & & \\ 0 & S_N & \cdots & S_N \end{bmatrix} \begin{bmatrix} a_1 \\ a_2 \\ \cdot \\ \cdot \\ \cdot \\ a_n \end{bmatrix} = \begin{bmatrix} 1 \\ 1 - \dfrac{p_2}{p_1} \\ \cdot \\ \cdot \\ 1 - \dfrac{p_N}{p_1} \end{bmatrix}$$

where, S_i's are the functions of p_i's and are obtained after above linear transformations. They are given by

$$S_i = \frac{p_1 - p_i}{(N-1)\, p_1}, \quad i = 1, 2, \ldots, N.$$

Thus the effective equations can be obtained as follows.

$$p_1 a_1 + \left(\frac{1-p_1}{N-1} \right) \sum_{r=2}^{n} a_r = 1 \tag{2.2.22}$$

and,

$$S_i \sum_{r=2}^{n} a_r = 1 - \frac{p_i}{p_1} = \frac{p_1 - p_i}{p_1}, \quad i = 2, \ldots, N$$

which when added over all i gives

$$\sum_{r=2}^{n} a_r = (N-1) \tag{2.2.23}$$

Substituting the value of $\sum_{r=2}^{n} a_r$ in Eq. (2.2.22) we get $a_1 = 1$, as $p_1 \neq 0$.

Thus we get the following two effective equations :

$$\left.\begin{aligned} a_r &= 1 \\ \sum_{r=2}^{n} a_r &= (N-1) \end{aligned}\right] \tag{2.2.24}$$

Now,

$$\text{Var}\left(\sum_{r=1}^{n} a_r X_r\right) = \sum_{r=1}^{n} a_r^2 \ \text{Var}(X_r) + \sum_{r \neq s}^{n} a_r a_s \ \text{Cov}(X_r, X_s)$$

From Lemma 2.2.1

$$V(X_1) = E(X_1)^2 - \{E(X_1)\}^2 = \sigma_1^2$$

where,

$$\sigma_1^2 = \sum_{i=1}^{N} p_i x_i^2 - \left(\sum_{i=1}^{N} p_i x_i\right)^2 \tag{2.2.25}$$

and for $r \geq 2$,

$$V(X_r) = E(X_r)^2 - \{E(X_r)\}^2$$

$$= \sigma_2^2$$

where,

$$\sigma_2^2 = \sum_{i=1}^{N} x_i^2 c_i - \left(\sum_{i=1}^{N} x_i c_i\right)^2 \tag{2.2.26}$$

with

$$c_i = p(X_r = x_i) = (1 - p_i)/(N-1)$$

Further,

$$\text{Cov}(X_1, X_r) = E(X_1 X_r) - E(X_1)\,E(X_r)$$

Now, for evaluating $E(X_1 X_r)$, we need probability of getting a specified unit selected at each of these draws. Let $P(X_1 = x_i, X_r = x_j)$ denote the probability of selecting i-th unit at the first draw and j-th unit at the r-th draw, then

$$P(X_1 = x_i; X_r = x_j) = P(X_1 = x_i).\, P(X_2 \neq x_j \mid X_1 = x_i).$$

$$P(X_3 \neq x_j \mid X_1 = x_i, X_2 \neq x_j), \ldots, P(X_{r-1} \neq x_j \mid X_1 = x_i, X_2 \neq x_j, \ldots, X_{r-2} \neq x_j).$$

$$P(X_r = x_j \mid X_1 = x_i, X_2 \neq x_j, \ldots, x_{r-1} \neq x_j)$$

$$= p_i \left(1 - \frac{1}{N-1}\right)\left(1 - \frac{1}{N-2}\right) \cdots \left(1 - \frac{1}{N-r+2}\right) \frac{1}{(N-r+1)}$$

$$= p_i \frac{N-2}{N-1} \frac{N-3}{N-2} \quad \cdots \quad \frac{N-r+1}{N-r+2} \cdot \frac{1}{N-r+1}$$

$$= \frac{p_i}{N-1} \quad \text{for any } r \geq 2. \tag{2.2.27}$$

Therefore,

$$E\,(X_1 X_r) = \sum_{i \neq j} x_i x_j \frac{p_i}{(N-1)}$$

$$= \frac{1}{N-1} \sum_{i \neq j} x_i p_i x_j$$

$$= \frac{1}{N-1}\left(T \sum_{i=1}^{N} x_i p_i - \sum_{i=1}^{N} x_i^2 p_i\right)$$

Thus,

$$\text{Cov}\,(X_1, X_r) = \frac{1}{N-1}\left(T \sum_{i=1}^{N} x_i p_i - \sum_{i=1}^{N} x_i^2 p_i\right) - \left(\sum_{i=1}^{N} x_i p_i\right)\left(\sum_{i=1}^{N} x_i c_i\right)$$

After some simplification

$$= -\frac{\sigma_1^2}{N-1}$$

$$= b_1 \text{ (say)} \tag{2.2.28}$$

Also we shall need $\text{Cov}\,(X_r, X_s)$, for $r, s \geq 2$ which can be evaluated on the same lines as discussed above, except the evaluation of probability which is given as below :

$P\,(X_r = x_i, X_s = x_j)$ $r, s \geq 2$ and $s \geq r$.

$= P\,(X_1 \neq x_i, x_j).\ P\,(X_2 \neq x_i, x_j \mid X_1 \neq x_i, x_j) \ldots$

$P\,(X_{r-1} \neq x_i, x_j \mid x_t \neq x_i, x_j$ for $t = 1, 2, \ldots, r-2)$.

$P\,(X_r = x_i \mid X_t \neq x_i, x_j$ for $t = 1, 2, \ldots, r-1)$.

$P\,(X_{r+1} \neq x_j \mid X_t \neq x_i, x_j$ for $t = 1, 2, \ldots, r-1; X_r = x_i) \ldots$

$P\,(X_{s-1} \neq x_j \mid X_t \neq x_i, x_j$ for $t = 1, 2, \ldots, r-1; X_r = x_i, X_{r+1} \neq x_j, \ldots, X_{s-2} \neq x_j)$.

$P\,(X_s = x_j \mid x_t \neq x_i, x_j$ for $t = 1, 2, \ldots, r-1; X_r = x_i, X_{r+1} \neq x_j, \ldots, X_{s-1} \neq x_j)$

$$= (1 - p_i - p_j)\frac{N-3}{N-1}\frac{N-4}{N-2}\frac{N-5}{N-3}\cdots\frac{N-r+1}{N-r+3}\frac{N-r}{N-r+2}$$

$$\frac{1}{N-r+1}\frac{N-r+1}{N-r}\frac{N-r+2}{N\ r-1}\cdots\frac{N-s+1}{N-s+2}\frac{1}{N-s+1}$$

$$= \frac{1-p_i-p_j}{(N-1)(N-2)} \quad \text{for } r, s \geq 2 \text{ and } s > r. \tag{2.2.29}$$

Therefore, $r \geq s$

$$\text{Cov}(X_r, X_s) = \sum_{i=1}^{N}\sum_{j(\neq i)}^{N} x_i x_j \frac{(1-p_i-p_j)}{(N-1)(N-2)} - (\Sigma\, x_i\, c_i)^2$$

$$= b_2 \text{ (say)} \tag{2.2.30}$$

Since a_1 is known, we choose $a_2, \ldots, a_n$ in such a way that the estimator is MVLUE, while the Eq. (2.2.4) is satisfied. For this let

$$\phi = a_1^2 \text{ Var}(X_1) + \sum_{r=2}^{n} a_r^2 \text{ Var}(X_r) + \sum_{r\neq s}^{n} a_r a_s \text{ Cov}(X_r, X_s) - \lambda\left(\sum_{r=2}^{n} a_r\right)$$

where, λ is the Lagrange's multiplier. The minimisation of ϕ is equivalent to minimising

$$\phi - a_1^2 \text{ Var}(X_1) = \sum_{r=2}^{n} a_r^2 \text{ Var}(X_r) + \sum_{r=2}^{n} a_1\, a_r \text{ Cov}(X_1, X_r)$$

$$+ \sum_{r=2}^{n}\sum_{s(\neq r)}^{n} a_r a_s \text{ Cov}(X_r, X_s) - \lambda\left(\sum_{r=2}^{n} a_r\right)$$

with respect to a_r for r =2, 3, . . ., n.

As $a_1 = 1$, now for $r \geq 2$.

$$\frac{\partial\phi}{\partial a_r} = 2a_r \text{ Var}(X_r) + \text{Cov}(X_1, X_r) + \sum_{s(\neq r)}^{n} a_s \text{ Cov}(X_r, X_s) - \lambda$$

equating to zero, we get

$$2a_r \text{ Var}(X_r) + \text{Cov}(X_1, X_r) + \sum_{s(\neq r)}^{n} a_s \text{ Cov}(X_r, X_s) = \lambda \tag{2.2.31}$$

Summing this equation over all values of r other than 1 we have

$$2\sum_{r=2}^{n} a_r \text{ Var}(X_r) + \sum_{r=2}^{n} \text{Cov}(X_1, X_r) + \sum_{r=2}^{n}\sum_{s(\neq r)}^{n} a_s \text{ Cov}(X_r, X_s)$$

$$-\lambda\,(n-1) = 0$$

or,

$$2\sigma_2^2\ (N-1) + (n-1)b_1 + b_2\ \sum_{r=2}^{n}\left(\sum_{s=2}^{n} a_s - a_r\right) - \lambda(n-1) = 0$$

or,

$$2\sigma_2^2\ (N-1) + (n-1)b_1 + b_2\ (N-1)(n-2) = \lambda(n-1)$$

which gives

$$\lambda = \frac{2\,\sigma_2^2\,(N-1)}{n-1} + b_1 + \frac{b_2}{n-1}(N-1)(n-2)$$

Substituting the value of λ in Eq. (2.2.31) we get

$$2a_r\,\sigma_2^2 + b_1 + \{(N-1) - a_r\}\,b_2 - 2\frac{(N-1)}{(n-1)}\sigma_2^2 - b_1 - b_2\frac{(N-1)(n-2)}{(n-1)} = 0$$

which gives

$$a_r = \frac{N-1}{n-1} \quad \text{for } r \geq 2 \tag{2.2.32}$$

Since $a_1 = 1$ from Eq. (2.2.24), we have MVLUE in this case as

$$\hat{T}_1 = X_1 + \frac{N-1}{n-1}\sum_{r=2}^{n} X_r \tag{2.2.33}$$

Also its variance is seen to be given by

$$V\left(\hat{T}_1\right) = \sigma_1^2 + \frac{(N-1)^2}{n-1}\{\,\sigma_2^2 + (n-2)\,b_2\,\} + 2\,(N-1)\,b_1$$

which after some simplification gives the desired result.

The reader may verify the first part of following Corollary (Prabhu Ajgaonkar, 1967), giving the unordered estimator, by using results, on unordering of Section 1.5, in Theorem 2.2.3. The second part of the Corollary giving the variance of the estimator is obtained directly.

Corollary 2.2.3

The estimator obtained by unordering the estimator in Eq. (2.2.33), is given by

$$\hat{T}_{cu}^{(l)} = \hat{T}_{uo}^{(p)}$$

$$= \frac{1}{\sum_{j=1}^{n} p_{i_j}} \left[\sum_{j=1}^{n} \left\{ p_{i_j} + \frac{N-1}{n-1} \left(\sum_{j' (\neq j)}^{n} p_{i_{j'}} \right) \right\} x_{i_j} \right]$$

The variance of the unordered estimator is

$$\sum_{l=1}^{\binom{N}{n}} \left(\hat{T}_{uo}^{(l)} \right) P(\zeta_l) - T^2$$

where,

$$P(\zeta_l) = P(S_p), \quad S_p \in \zeta_l$$

$$= \sum_{t=1}^{n!} P(s_t), \quad s_t \in S_p$$

Putting $p_i = 1/N$, we get the results for SRSWOR scheme.

The reader may further verify that the relative efficiency as given by Prabhu Ajgaonkar (1967) of the unordered estimator compared to the ordered estimator is 288 for the following population :

Unit	p	x
1	.1	0.2
2	.2	0.6
3	.3	0.9
4	.4	0.8

We now present the result for Midzuno scheme of sampling with replacement in the following theorem.

Theorem 2.2.4

The MVLUE exists in T_1-class for Midzuno scheme of sampling with replacement and is given by

$$\hat{T}_1 = \frac{N}{n-1} \sum_{r=2}^{n} X_r$$

and its variance is given by

$$V\left(\hat{T}_1\right) = \frac{N^2}{n-1} \frac{N-1}{N} S^2$$

Proof :

Under the said Sampling Scheme the probability matrix is given by

$$P = \begin{bmatrix} p_1 & \frac{1}{N} & \cdots & \frac{1}{N} \\ p_2 & \frac{1}{N} & \cdots & \frac{1}{N} \\ . & & & \\ . & & & \\ p_N & \frac{1}{N} & \cdots & \frac{1}{N} \end{bmatrix} \qquad (2.2.34)$$

For the estimator to be unbiased we must have

$$\begin{bmatrix} p_1 & \frac{1}{N} \cdots & \frac{1}{N} \\ p_2 & \frac{1}{N} \cdots & \frac{1}{N} \\ . & & \\ . & & \\ . & & \\ p_N & \frac{1}{N} \cdots & \frac{1}{N} \end{bmatrix} \begin{bmatrix} a_1 \\ a_2 \\ . \\ . \\ . \\ a_n \end{bmatrix} = \begin{bmatrix} 1 \\ 1 \\ . \\ . \\ . \\ 1 \end{bmatrix} \qquad (2.2.35)$$

Assuming without loss of generality, $p_1 \neq p_2$ as in the proof of Theorem 2.2.3, it can be easily verified that the ranks of P in (2.2.34) and of P' obtained by adding a column of 1's in (2.2.34) are same and are equal to 2. Therefore, there are only two effective equations. We proceed to find these equations. Subtracting first row from the second to N-th rows in Eq. (2.2.35), we have

$$\begin{bmatrix} p_1 & \frac{1}{N} & \cdots & \frac{1}{N} \\ p_2 - p_1 & 0 & \cdots & 0 \\ & & & \\ & & & \\ p_N - p_1 & 0 & \cdots & 0 \end{bmatrix} \begin{bmatrix} a_1 \\ a_2 \\ . \\ . \\ . \\ a_n \end{bmatrix} = \begin{bmatrix} 1 \\ 0 \\ . \\ . \\ . \\ 0 \end{bmatrix} \qquad (2.2.36)$$

which is equivalent to

$$\left.\begin{array}{lll} p_1 a_1 + \dfrac{1}{N} \sum\limits_{r=2}^{n} a_r = 1 & & \ldots (1) \\ (p_2 - p_1)\, a_1 = 0 & & \ldots (2) \\ \cdot & & \\ \cdot & & \\ \cdot & & \\ (p_N - p_1) a_1 = 0 & & \ldots (N) \end{array}\right] \quad (2.2.37)$$

Since $(p_2 - p_1) \neq 0,\ a_1 = 0$. from Eq. (2) of (2.2.37) and its Eqs. (3) to (N) do not give any new result. Therefore, we have the following two effective equations :

$$\left.\begin{array}{ll} a_1 p_1 + \dfrac{1}{N} \sum\limits_{r=2}^{n} a_r = 1 & \text{(i)} \\ a_1 = 0 & \text{(ii)} \end{array}\right] \quad (2.2.38)$$

clearly from (i) and (ii) of (2.2.38) we have

$$\sum_{r=2}^{n} a_r = N \quad (2.2.39)$$

Further we have

$$V\left(\hat{T}_1\right) = V\left(\sum_{r=1}^{n} a_r X_r\right)$$

$$= \sum_{r=1}^{n} a_r^2 \operatorname{Var}(X_r). \quad (2.2.40)$$

As, the covariance terms are zero.

Here,

$$V(X_1) = E(X_1)^2 - \{E(X_1)\}^2$$

$$= \sum_{i=1}^{N} x_i^2 p_i - \left(\sum_{i=1}^{N} x_i p_i\right)^2 = \sigma_1^2 \text{ (say)}$$

and

$$V(X_r) = \frac{1}{N}\left\{\sum_{i=1}^{N} x_i^2 - \frac{(\Sigma x_i)^2}{N}\right\} = \frac{N-1}{N} S^2 \ , \ r \geq 2$$

where, S^2 is defined in Theorem 2.2.1.

We obtain below the condition, under which Midzuno's Estimator for population total is MVLUE. Let

$$\phi = \sum_{r=2}^{n} a_r^2 \, Var(X_r) - 2\lambda \left(\sum_{r=2}^{n} a_r \right)$$

Differentiating with respect to a_r for $r \geq 2$, and equating this to zero gives,

$$\lambda = a_r \, V(X_r)$$

Summing the above equation over $r \geq 2$, we get

$$a_r = \left(\frac{N}{n-1} \right) \tag{2.2.41}$$

The proof is completed by substituting the values of a_r for r = 1, 2, . . ., n in T_1-class estimator and in variance formula (2.2.40).

From Theorems 2.2.2 and 2.2.4, it is clear that the esimator for SRSWR is always more efficient than the estimator for Midzuno sampling scheme with replacement, which is a varying probability sampling scheme. A detailed comparison by Tikkiwal (1966) suggested necessary caution in using varying probability sampling schemes in preference to equal probability sampling schemes.

The reader may now verify the first part of the following Corollary using results of Section 1.4 in Theorem 2.2.4. The first part of the Corollary then gives the classical estimator for SRSWR scheme by putting $p_i = 1/N$ for all i. The variance of the estimator in second part of the Corollary is directly obtained.

Corollary 2.2.4

The estimator obtained by unordering the estimator in Theorem 2.2.4 is given by

$$\hat{T}_{uo}^{(p)} = \frac{N}{n-1} \left\{ \sum_{j=1}^{k} x_{i_j} \, R_{i_j}^{(p)} - \frac{\sum_{j=1}^{k} x_{i_j} \, R_{i_j}^{(p)} \, p_{i_j}}{\sum_{j=1}^{k} p_{i_j} \, R_{i_j}^{(p)}} \right\}$$

The variance of this estimator is

$$\sum_{k=1}^{n} \sum_{l=1}^{\binom{N}{k}} \sum_{p=1}^{M_{kl}} \left(\hat{T}_{uo}^{(p)} \right)^2 P(S_p) - T^2$$

where,

$$P(S_p) = \frac{M_{klp}}{n.N^{n-1}} \sum_{j=1}^{k} p_{i_j} \, R_{i_j}^{(p)}$$

We have considered so far Midzuno sampling scheme for sampling with or without replacement; when the units are drawn with unequal probabilities at the first draw, but at the subsequent draws they are drawn with equal probabilities. There is obvious generalisations to these sampling schemes.

Prabhu Ajagaonkar (1962) has discussed in detail these extensions. For sampling without replacement, when the units at the first two draws are drawn by Horvitz and Thomspon's sampling schemes (discussed later in this chapter) and at the subsequent draws by SRSWOR scheme; he obtains the estimator as

$$\hat{T}_1 = ax_1 + bx_2 + \frac{c}{n-2} \sum_{r=3}^{n} X_r \tag{2.2.42}$$

However, when sampling without replacement schemes involving unequal probabilities are used on more than two draws; no results are available as the derivations get complicated.

For sampling with replacement, if $p_{ir'} = p_i$ for all i and $r' = 1, 2, \ldots, r$ and $p_{ir'} = 1/N$ for all i and $r' > r$; then he obtains the estimator as

$$\hat{T}_1 = \frac{N}{n-r} \sum_{r'=r+1}^{n} X_{r'} \tag{2.2.43}$$

and its variance as

$$V\left(\hat{T}_1\right) = \frac{N^2}{n-r} \cdot \frac{N-1}{N} S^2 \tag{2.2.44}$$

In order to improve their efficiency, these estimators can be unordered as in the special case. For sampling with replacement, in fact we can improve their efficiency further by obtaining unbiased unordered estimators as discussed in part (b) of Chapter 1.

So far we have considered the sampling schemes, where, T_1-class is nonempty. Now we present a varying probability sampling scheme without replacement, where T_1-class is empty in the sense that their does not exist any unbiased estimator in T_1-class for this sampling scheme.

For n = 2, let p (i, 1) = p_i for all i

and,

$$p\{(j,2)\,|\,(i,1)\} = \begin{cases} \dfrac{p_j}{1-p_i} & , \ j \neq i \\ 0 & , \ j = i \end{cases}$$

This scheme is given by Horvitz and Thompson (1952). When $p_i \; \alpha \; z_i$ the size measure of i-th units, this scheme reduces to the well known probability proportional to size (pps) without replacement scheme.

Theorem 2.2.5

The T_1-class is empty for Horvitz-Thompson sampling scheme.

Proof :

Here

$$P_{j2} = \sum_{i\,(\neq j)}^{N} p\,\{(j,2)\,|\,(i,1)\}\; p\,(i,1)$$

$$= \sum_{i(\neq j)}^{N} \frac{p_i\, p_j}{(1-p_i)}$$

$$= p_j \left(S - \frac{p_j}{1-p_j} \right) \qquad (2.2.45)$$

where,

$$S = \sum_{i=1}^{N} \left(\frac{p_i}{1-p_i} \right)$$

In a varying probability sampling scheme, atleast two p_i's are not equal. Let there be p_1 and p_2, without any loss of generality. In order that at the second draw, at least two conditional probabilities are not equal. It is necessary that we assume that a third p_i is not equal to either p_1 or p_2. Let this be p_3. Thus, we assume that $p_1 \neq p_2 \neq p_3$. The probability matrix P will have rank 2, if

$$\begin{vmatrix} p_1 & \left(S - \dfrac{p_1}{1-p_1}\right) p_1 \\ p_2 & \left(S - \dfrac{p_2}{1-p_2}\right) p_2 \end{vmatrix} \neq 0$$

After simplification,

$$p_1 p_2 \left(\frac{p_1}{1-p_1} - \frac{p_2}{1-p_2}\right) \neq 0.$$

which is so as $p_1 \neq p_2$.

The rank of P' will be three, if

$$\begin{vmatrix} p_1 & \left(S - \dfrac{p_1}{1-p_1}\right) p_1 & 1 \\ p_2 & \left(S - \dfrac{p_2}{1-p_2}\right) p_2 & 1 \\ p_3 & \left(S - \dfrac{p_3}{1-p_3}\right) p_3 & 1 \end{vmatrix} \neq 0$$

Subtrating the first row from the second and third rows and then simplifying, the rank of P' will be three, if

$$\frac{(p_1 - p_2)(p_2 - p_3)(p_3 - p_1)}{(1 - p_1)(1 - p_2)(1 - p_3)} \neq 0$$

which is so as $p_1 \neq p_2 \neq p_3$. By Lemma 2.2.1 there does not exist any unbiased estimator in T_1-class for this sampling scheme.

We state the following Theorem for varying probability scheme with replacement in which $p_{ir} = p_i$ for all r and i. Thus,

$$P = \begin{bmatrix} p_1 & \cdots & p_1 \\ p_2 & \cdots & p_2 \\ \cdot & & \\ \cdot & & \\ \cdot & & \\ p_N & \cdots & p_N \end{bmatrix}$$

in this case.

Theorem 2.2.6

The T_1-class is empty for varying probability sampling scheme with replacement.

Proof :

This Theorem follows by noting that R(P) = 1 and R(P')=2.

EXERCISES

2.1. A sample of size n is drawn with simple random sampling and without replacement from a population of size N. Let

$$\hat{T}_n = \sum_{r=1}^{n} \beta_r X_r \qquad (1)$$

be any linear estimator of the population mean $\bar{x}_N$, obtained by associating weight β_r to X_r, the observation on the unit drawn at the r-th draw for r= 1,2, . . ., n and then taking the sum. Then show that:

(a) $\hat{T}_n$ is an unbiased estimator of $\bar{x}_N$ if and only if

$$\sum_{r=1}^{n} \beta_r = 1 \qquad (2)$$

(b) Under condition (2) the variance of $\hat{T}_n$ is given by

$$V\left(\hat{T}_n\right) = \frac{S^2}{N}\left(N \sum_{r=1}^{n} \beta_r^2 - 1\right) \qquad (3)$$

where S^2 is mean square for the population.

(c) $V\left(\hat{T}_n\right)$ is minimized subject to condition (2) if $\beta_r = 1/n$ for r = 1, 2, . . ., n. Hence prove that the sample mean $\overline{X}_n$ is the minimum variance unbiased estimator of $\bar{x}_N$ in the class of linear estimators given by (1).

2.2 Show that MVLUE of population total T exists in T_1-class for Midzuno scheme of sampling without replacement and it is given by

$$\hat{T}_1 = X_1 + \frac{N-1}{n-1} \sum_{r=2}^{n} X_r$$

where, X_r $(r=1,\ldots,n)$ denotes the r-th draw. Also obtain the expression of variance of $\hat{T}_1$.

2.3. Show that MVLUE of population total T exists in T_1-class for Midzuno scheme of sampling with replacement and is given by

$$\hat{T}_1 = \frac{N}{n-1}\sum_{r=2}^{n} X_r$$

and the variance is given by

$$V\left(\hat{T}_1\right) = \frac{N(N-1)}{(n-1)} S^2$$

where, $$S^2 = \frac{1}{N-1}\sum_{i=1}^{N}(x_i - \bar{x}_N)^2.$$

Is $\hat{T}_1$ is more efficient than the estimator $\hat{T} = N\bar{X}_n$ under SRSWR ?

2.4 Show that MVLUE does not exist in T_1-class of linear estimators under Horvitz - Thompson's sampling scheme.

2.5 Using Murthy's technique of unordering, obtain an unordered estimator $\hat{T}_{uo}$ corresponding to the ordered estimator $\hat{T}_1$ given in Exercise 2.2, and obtain expression of variance of $\hat{T}_{uo}$. For the following population of size N=4 and sample size n=2 calculate the efficiency of $\hat{T}_{uo}$ with respect to $\hat{T}_1$.

Unit	p	x
1	0.1	0.2
2	0.2	0.6
3	0.3	0.9
4	0.4	0.8

Appendix I

The abstract of the paper entitled - On Horvitz and Thompson's T-class Estimators (Preliminary report). S.G. Prabhu Ajgaonkar and B.D. Tikkiwal, Karnatak University, Dharwar, India (By title) Ann. Math. Statist. Vol. 32, No. 3. 1961, p. 923.

Horvitz and Thompson (J.A.S.A., 1952), while discussing sampling with varying probability and without replacement, have given three classes of estimators. If an empty class is that where the unbiased estimators independent of population values do not exist, it is shown that their T_1-class is in general an empty class when sampling with varying probability is adopted. However, when Midzuno's system of sampling is adopted with replacement, such a class of estimators exists and has a minimum variance unbiased estimator in the class independent of the population values. The T_3-class, which is non-empty, has no minimum variance unbiased estimator independent of the population values even when simple random sampling is adopted. The non-empty T_2-class is known to have only one unbiased estimator and so is a minimum variance unbiased estimator. It is noted that for Midzuno's system of sampling with replacement T_2-class estimator has a smaller variance than that of the minimum variance unbiased estimator of T_1-class. It is further noted for sampling with varying probability and with replacement that the minimum variance unbiased estimator in the over all class consisting of the classes T_1 and T_2 is the unbiased estimator in the T_2 class. However, the relative efficiency of T_2-class estimator and an estimator in T_3-class depends upon the probability system adopted.

CHAPTER 3

T_2 - Class Estimators

3.1 INTRODUCTION

This chapter gives in Section 3.2 a unified general discussion, of T_2-class estimators for sampling with varying probabilities and without replacement, due to Prabhu Ajgaonkar and Tikkiwal (1961), the generalised T_2-class itself being defined earlier by Tikkiwal (1960). The estimator in T_2-class (Horvitz and Thompson, 1952) and its variance for sampling without replacement is obtained as a special case in Sub-section 3.3.1. Sub-section 3.3.2 gives three estimators of the variance and a discussion of their relative merits. In Sub-section 3.3.3 non-negative variance estimators are discussed. In Sub-section 3.3.4 certain probabilities of inclusion; of the population units in singles and pairs, required in the estimator and in its variance as well as in the estimator of the variance for sampling without replacement; are calculated. Section 3.4 gives the estimator and its variance together with the estimator of the variance for sampling with replacement. The relative efficiency of the two sampling systems with and without replacements is also discussed in the same Section. Section 3.5 gives the results, for sampling without replacement, due to Rao, Hartley and Cochran (1962) to overcome the difficulty in calculation of the probabilities in general for $n > 2$. Sections 3.6 and 3.7 give the general theory of multi-stage sampling with varying probabilities and with or without replacement and a comparison of these two sampling schemes. It is shown (Tikkiwal, 1965) that the various estimators due to Sukhatme and Narain, Durbin and Ecimovic belongs to the extended T_2-class for multi-stage designs. We shall refer this class of estimators simply as T_2-class, as no confusion is likely to arise because of this.

3.2 A GENERAL DISCUSSION OF T_2-CLASS ESTIMATORS

The problem faced in surveys is one that of estimation of totals (or populations means) for different characteristics of the population at a given time or at different times. The estimators generally adopted are linear functions of the sample observations on the units in the sample drawn from the population, as discussed in Chapter 1. The linear functions, which are obtained by associating with a particular unit a weight dependent on the unit itself give rise to estimators in T_2-class. In what follows, we confine ourselves to the problem of estimation of T, the total for a particular characteristic of the population, in T_2-class.

When a sample of size n is drawn from the population with varying probabilities and with or without replacement in order to estimate the total T, an estimator $\hat{T}_2$ of T in T_2-class, in general is given by

$$\hat{T}_2 = \sum_{r=1}^{n} \beta_{ro}^{s_t} X_r \tag{3.2.1}$$

where X_r denotes the random outcome at the r-th draw and $\beta_{ro}^{s_t}$ denotes the weight to be associated with X_r for a given sample s_t. An estimator in T_2-class is obtained by taking $\beta_{ro}^{s_t} = \beta_i$ if $X_r = x_i$, for all t, x_i denoting the value of the i-th unit in the population for i=1, 2,. . ., N. Let p_{ir} denote the probability of i-th unit of the population occurring at the r-th draw. For an estiamtor in T_2-class,

$$E\left(\hat{T}_2\right) = \sum_{r=1}^{n} \sum_{i=1}^{N} x_i \beta_i \, p_{ir}$$

$$= \sum_{i=1}^{N} x_i \beta_i \left(\sum_{r=1}^{n} p_{ir} \right)$$

$$= \sum_{i=1}^{N} x_i \beta_i \pi_i \, ; \tag{3.2.2}$$

In order that $\hat{T}_2$ be an unbiased estimator; we must have $\beta_i = 1/\pi_i$, a unique value for β_i, for all i=1,2,. . ., N. Thus there is only one unbiased estimator in T_2-class and therefore, it is also the minimum variance linear unbiased estimator (MVLUE) in T_2-class. Now,

$$V\left(\hat{T}_2\right) = E\left(\sum_{r=1}^{n} \beta_{ro}^{s_t} X_r \right)^2$$

$$= E\left\{ \sum_{r=1}^{n} \left(\beta_{ro}^{s_t} X_r \right)^2 + \sum_{r \neq s}^{n} \beta_{ro}^{s_t} \beta_{so}^{s_t} X_r X_s \right\} - T^2$$

$$= \sum_{r=1}^{n} \sum_{i=1}^{N} \frac{x_i^2}{\pi_i^2} p_{ir} + \sum_{r \neq s}^{n} \sum_{i=1}^{N} \sum_{j \neq i}^{N} \frac{x_i x_j}{\pi_i \pi_j} p(ir, js) - T^2$$

$$= \sum_{i=1}^{N} \frac{x_i^2}{\pi_i^2} \sum_{r=1}^{n} p_{ir} + \sum_{i=1}^{N} \sum_{j \neq i}^{N} \frac{x_i}{\pi_i} \frac{x_j}{\pi_j} \sum_{r \neq s}^{n} p\ (ir,\ js) - T^2$$

$$= \sum_{i=1}^{N} \frac{x_i^2}{\pi_i} + \sum_{i=1}^{N} \sum_{j \neq i}^{N} \frac{x_i}{\pi_i} \frac{x_j}{\pi_j} \pi_{ij} - T^2 \tag{3.2.3}$$

We note from the proof of results given in Eqs. (3.2.2) and (3.2.3) easily the following general results (Kendal and Stuart, Vol. 3, 1976, third edition, Eq. 39.19, p. 178 and Eq. 39.33, p. 182).

$$E\left\{ \sum_{r=1}^{n} g(X_r) \right\} = \sum_{i=1}^{N} \{\pi_i\, g(x_i)\} \tag{3.2.4}$$

and,

$$E\left\{ \sum_{r=1}^{n} \sum_{s(\neq r)}^{n} g(X_r, X_s) \right\} = \sum_{i=1}^{N} \sum_{j(\neq i)}^{N} \pi_{ij}\ g\ (x_i, x_j) \tag{3.2.5}$$

It may be noted that, in case of sampling without replacement, the terms π_i and π_{ij} denote respectively probability of inclusion of i-th unit in the sample and the probability of inclusion of i-th and j-th units in the sample, where as in case of sampling with replacement they provide upper limits to these probabilities.

We state the main results below in the form of a theorem.

Theorem 3.2.1 (Prabhu Ajgaonkar and Tikkiwal, 1961)

When a sample of size n is drawn from the population with varying probabilities and with or without replacement in order to estimate the population total T, then its only unbiased estimator in T_2- class is $\hat{T}_2$ given by Eq. (3.2.1). The variance of this estimator is given by Eq. (3.2.3).

3.3 SAMPLING WITH VARYING PROBABILITIES AND WITHOUT REPLACEMENT

We present the results for sampling with varying probabilities and without replacement in four parts. The first part gives the variance of Horvitz and Thompson estimator, the second part the three unbiased estimators of variance, third part discusses the non-negativity of different estimators of the variance, and the fourth part gives the calculation of certain probabilities of inclusion required in estimating the variance.

3.3.1 The Variance of Horvitz-Thompson Estimator

Let $\hat{T}_{ht}$ denote the estimator for sampling without replacement. Since $p(ir, js) = 0$ for $i = j$, from Theorem 3.2.1 we have,

$$V\left(\hat{T}_{ht}\right) = E\left(\hat{T}_{ht}\right)^2 - \left\{E\left(\hat{T}_{ht}\right)\right\}^2$$

$$= \sum_{i=1}^{N} \frac{x_i^2}{\pi_i} + \sum_{i=1}^{N} \sum_{j(\neq i)=1}^{N} \frac{x_i}{\pi_i} \frac{x_j}{\pi_j} \pi_{ij} - \left(\sum_{i=1}^{N} x_i\right)^2$$

$$= \sum_{i=1}^{N} \frac{x_i^2}{\pi_i} + \sum_{i=1}^{N} \sum_{j(\neq i)=1}^{N} \frac{x_i}{\pi_i} \frac{x_j}{\pi_j} \pi_{ij} - \sum_{i=1}^{N} x_i^2 - \sum_{i=1}^{N} \sum_{j(\neq i)=1}^{N} x_i x_j$$

$$= \sum_{i=1}^{N} \frac{x_i^2}{\pi_i} (1 - \pi_i) + \sum_{i=1}^{N} \sum_{j(\neq i)=1}^{N} \frac{x_i x_j}{\pi_i \pi_j} (\pi_{ij} - \pi_i \pi_j) \tag{3.3.1}$$

A result, in the present form, due to Horvitz and Thompson (1952, Eq. 9, p. 670). these results for sampling without replacement can be obtained with slight modification in the estimator and its variance given by Narain (1951).

Note: It may be noted that if $\pi_i \, \alpha \, x_i \; (i = 1, \ldots, N)$, $V\left(\hat{T}_{ht}\right) = 0$. This is called the ratio estimator property of the Horvitz-Thompson estimator. In pratice, x-values will be unknown and instead, values of some auxiliary variables y closely related to the main variable x may be known. In such cases it may be assumed that y is approximately proportional to x.

A fixed effective design of size n [FS(n)] with $\pi_i \, \alpha \, p_i \; (= y_i / Y)$ is called a π ps design or IPPS (inclusion probability proportional to size) design. Since $\pi_i \leq 1$ it is required that $y_i \leq Y/n \; \forall i$. One may, therefore, expect that $V\left(\hat{T}_{ht}\right)$ based on π ps design small, provided auxiliary variable y is suitably chosen.

3.3.2 The Three Unbiased Estimators of the Variance of $\hat{T}_{ht}$ and Their Relative Merits

We present three unbiased estimators of the variance of $\hat{T}_{ht}$ as given in Eq. (3.3.1) and discuss their relative merits. The unbiased estimator, of the variance of $\hat{T}_{ht}$, due to Horvitz and Thompson is

$$\hat{V}_{ht} = \hat{T}^2_{ht} - (\text{an unbiased estimator of } T^2)$$

$$= \hat{T}^2_{ht} - \sum_{r=1}^{n} X_r^2 \beta_{ro}^{st} - \sum_{r \neq s}^{n} X_{r,s} \tag{3.3.2}$$

where, $\beta_{ro}^{st} = 1/\pi_i$ if $X_r = x_i$ and

$X_{r,s} = \dfrac{x_i\, x_j}{\pi_{ij}}$ if x_i and x_j units occur at the r-th and s-th draws respectively. It is to be noted here that this estimator of $V(\hat{T}_{ht})$ is unbiased but it assumes negative values, as shown by Sen (1953 a), Yates and Grundy (1953).

In order to obtain the alternative unbiased estimator of the variance of $\hat{T}_{ht}$ due to Sen, Yates and Grundy, we first present the results due to Horvitz and Thompson (1952) in the form of following Lemma.

Lemma 3.3.1

(i) $\sum_{i=1}^{N} \pi_i = n$

(ii) $\sum_{j(\neq i)}^{N} \pi_{ij} = (n-1)\,\pi_i$

Proof :

(i) $\sum_{i=1}^{N} \pi_i = \sum_{i=1}^{N} \sum_{r=1}^{n} p_{ir} = \sum_{r=1}^{n} \left(\sum_{i=1}^{N} p_{ir} \right) = n$

noting that $\sum_{i=1}^{N} p_{ir} = 1$ for any given r.

(ii) Let $p(js|ir)$ denote the conditional probability of j-th unit occurring at s-th draw given that i-th unit has occurred at the r-th draw. Then

$$\sum_{j(\neq i)}^{N} \pi_{ij} = \sum_{j(\neq i)}^{N} \sum_{r=1}^{n} p_{ir} \sum_{s(\neq r)}^{n} \{p(js|ir)\}$$

$$= \sum_{r=1}^{n} p_{ir} \left[\sum_{s(\neq r)}^{n} \sum_{j(\neq i)}^{N} \{p(js|ir)\} \right]$$

$$= \sum_{r=1}^{n} p_{ir} (n-1) = (n-1)\, \pi_i$$

Corollary 3.3.1

$$\sum_{j(\neq i)}^{N} (\pi_{ij} - \pi_i \pi_j) = (n-1)\, \pi_i - \pi_i \sum_{j(\neq i)} \pi_j$$

$$= (n-1)\, \pi_i - \pi_i\, (n - \pi_i)$$

$$= \pi_i^2 - \pi_i$$

Using the above Corollary in Eq. (3.3.1)

$$V(\hat{T}_{ht}) = \sum_{i=1}^{N} \left[\frac{x_i^2}{\pi_i^2} \left\{ \sum_{j(\neq i)}^{N} (-\pi_{ij} + \pi_i \pi_j) \right\} + \sum_{i \neq j}^{N} \left\{ \frac{x_i x_j}{\pi_i \pi_j} (\pi_{ij} - \pi_i \pi_j) \right\} \right]$$

$$V(\hat{T}_{ht}) = \left[\frac{x_1^2}{\pi_1^2} \{(-\pi_{12} + \pi_1 \pi_2) + (-\pi_{13} + \pi_1 \pi_3) + \ldots + (-\pi_{1N} + \pi_1 \pi_N)\} \right.$$

$$+ \frac{x_2^2}{\pi_2^2} \{(-\pi_{21} + \pi_2 \pi_1) + (-\pi_{23} + \pi_2 \pi_3) + \ldots + (-\pi_{2N} + \pi_2 \pi_N)\}$$

$$+ \ldots$$

$$\cdot$$

$$\cdot$$

$$\left. + \frac{x_N^2}{\pi_N^2} \{(-\pi_{N1} + \pi_N \pi_1) + (-\pi_{N2} + \pi_N \pi_2) + \ldots + (-\pi_{N(N-1)} + \pi_N \pi_{N-1})\} \right]$$

$$+ \frac{2x_1 x_2}{\pi_1 \pi_2}(\pi_{12} - \pi_1 \pi_2) + \ldots + \frac{2x_1 x_N}{\pi_1 \pi_N}(\pi_{1N} - \pi_1 \pi_N)$$

$$+ \frac{2x_2 x_3}{\pi_2 \pi_3}(\pi_{23} - \pi_2 \pi_3) + \ldots + \frac{2x_2 x_N}{\pi_2 \pi_N}(\pi_{2N} - \pi_2 \pi_N)$$

$$\vdots$$

$$+ \frac{2x_N x_{N-1}}{\pi_N \pi_{N-1}}(\pi_{N(N-1)} - \pi_N \pi_{N-1})$$

$$= \left[\frac{x_1^2}{\pi_1^2}(\pi_1\pi_2 - \pi_{12}) + \frac{x_2^2}{\pi_2^2}(\pi_1\pi_2 - \pi_{12}) - \frac{2x_1 x_2}{\pi_1 \pi_2}(\pi_1\pi_2 - \pi_{12}) \right.$$

$$+ \ldots$$

$$\vdots$$

$$+ \frac{x_1^2}{\pi_1^2}(\pi_1\pi_N - \pi_{1N}) + \frac{x_N^2}{\pi_N^2}(\pi_1\pi_N - \pi_{1N}) - \frac{2x_1 x_N}{\pi_1 \pi_N}(\pi_1\pi_N - \pi_{1N})$$

$$+ \frac{x_2^2}{\pi_2^2}(\pi_2\pi_3 - \pi_{23}) + \frac{x_3^2}{\pi_3^2}(\pi_2\pi_3 - \pi_{23}) - \frac{2x_2 x_3}{\pi_2 \pi_3}(\pi_2\pi_3 - \pi_{23})$$

$$+ \frac{x_2^2}{\pi_2^2}(\pi_2\pi_N - \pi_{2N}) + \frac{x_N^2}{\pi_N^2}(\pi_2\pi_N - \pi_{2N}) - \frac{2x_2 x_N}{\pi_2 \pi_N}(\pi_2\pi_N - \pi_{2N})$$

$$+ \ldots$$

$$\vdots$$

$$+ \frac{x_{N-1}^2}{\pi_{N-1}^2}(\pi_{N-1}\pi_N - \pi_{N-1,N}) + \frac{x_N^2}{\pi_N^2}(\pi_{N-1}\pi_N - \pi_{N-1,N})$$

$$\left. - \frac{2x_{N-1} x_N}{\pi_{N-1} \pi_N}(\pi_{N-1}\pi_N - \pi_{N-1,N}) \right]$$

$$= \sum_{i=1}^{N} \sum_{j>i}^{N} \left(\frac{x_i}{\pi_i} - \frac{x_j}{\pi_j} \right)^2 (\pi_i \pi_j - \pi_{ij}) \tag{3.3.3}$$

Therefore, an alternate unbiased estimator due to Sen, Yates and Grundy by Eq. (3.2.5) is

$$\hat{V}_{syg} = \sum_{r=1}^{n} \sum_{s>r}^{n} C_{rs} \tag{3.3.4}$$

where,

$$C_{rs} = (\pi_i \, \pi_j - \pi_{ij}) / \pi_{ij} \, . (x_i / \pi_i - x_j / \pi_j)^2 \tag{3.3.5}$$

if, at the r-th and s-th draws, i-th and j-th units occur in respective order.

3.3.3 Non-negative Variance Estimators

Here $\hat{V}_{ht}$ is known to be unbiased provided unbiased estimation of $V(\hat{T}_{ht})$ is possible at all, i.e. provided $\pi_{ij} > 0$ for all i and j, while $\hat{V}_{syg}$ is known to be unbiased provided the sampling design p is of fixed size and has $\pi_{ij} > 0$ for all i and j.

Let us finally remark that the condition

$$\pi_{ij} \le \pi_i \, \pi_j \qquad \text{for all i and j}$$

is clearly sufficient for $\hat{V}_{syg}$ to be non-negative.

For estimators of the variance of $\hat{T}_{ht}$, several suggestions have been given. Of the two classical ones, $\hat{V}_{ht}$ and $\hat{V}_{syg}$ the former was given by Horvitz and Thompson (1952) and the latter by Yates and Grundy (1953) and by Sen (1953); further suggestions by Prabhu-Ajgaonkar (1967 a) and Ramakrishnan (1970).

The relative merits of $\hat{V}_{ht}$ and $\hat{V}_{syg}$ can be described as follows. In sampling designs which are not of fixed size, $\hat{V}_{syg}$ is biased and thus sometimes considered unsuitable. On the other hand, if the design is of fixed size, one generally prefers $\hat{V}_{syg}$ to $\hat{V}_{ht}$ on the ground that the former is believed to show smaller sampling fluctuations; in particular $\hat{V}_{syg}$ is believed to take negative values `less often' than $\hat{V}_{ht}$. For some sampling designs $\hat{V}_{syg}$ is even known never to take negative values; see Sen (1953), Raj (1956), Rao (1961), Rao (1963), Rao (1965), Seth (1966) and Lanke (1975).

Yates and Grundy (1953) in their pseudo empirical study (Table 3, p. 258) and also in the study of Horvitz and Thompson's population of 20 city blocks noted that Horvitz and Thompson's estimator gives negative values where their estimator does not. Durbin (1953, p. 264), showed that Sen, Yates and Grundy's estimator can also assume negative values. Sen (1953 a, Theorems 1 and 2) showed that this estimator of variance is always positive, when sampling is done according to Midzuno's scheme for any n described by Horvitz and Thompson as sampling scheme 1 and also according to Horvitz and Thompson's scheme 2, when n = 2 (1952, pp. 678-679). Hartley and Rao (1962, p. 364) proved that the result for Horvitz and Thompson's scheme 2 under restrictive conditions perhaps being unaware of the result due to Sen. These studies (cf. some exercises based on these studies) indicate that Sen, Yates and Grundy's estimator is likely to assume negative values much less often than Horvitz and Thompson's estimator.

Sen (1953) suggested a third estimator $\hat{V}_s$, which is essentially the same as Horvitz and Thompson's estimator $\hat{V}_{ht}$ where it assumes positive values with the difference that for negative values of the latter estimator, it takes the value zero. Thus,

$$\hat{V}_s = \begin{cases} \hat{V}_{ht}, & \hat{V}_{ht} > 0 \\ 0, & \hat{V}_{ht} \le 0 \end{cases} \tag{3.3.6}$$

Further,

$$E\left(\hat{V}_{ht}\right) = \Sigma_1 \; p(s_t)\left(\hat{V}_{ht}\right) + \Sigma_2 \; p(s_t)\left(\hat{V}_{ht}\right)$$

where, Σ_1 is the summation over those s_t's for which $\left(\hat{V}_{ht}\right)$ is positive and Σ_2 is the summation over all the remaining s_t's for which $\hat{V}_{ht}$ is neative.

Therefore,

$$E\left(\hat{V}_{ht}\right) = E\left(\hat{V}_s\right) - k \quad \text{where, } k \geq 0.$$

Thus,

$$E\left(\hat{V}_s\right) = E\left(\hat{V}_{ht}\right) + k \tag{3.3.7}$$

We note that $\hat{V}_s$ is a positively biased estimator of $V\left(\hat{T}_{ht}\right)$.

Also,

$$E\left\{\hat{V}_{ht} - V\left(\hat{T}_{ht}\right)\right\}^2 = \Sigma_1 \; p(s_t)\left\{\hat{V}_{ht} - V\left(\hat{T}_{ht}\right)\right\}^2 + \Sigma_2 \; p(s_t)\left\{\hat{V}_{ht} - V\left(\hat{T}_{ht}\right)\right\}^2$$

$$= \Sigma_1 \; p(s_t)\left\{\hat{V}_{ht} - V\left(\hat{T}_{ht}\right)\right\}^2$$

$$+ \Sigma_2 \; p(s_t)\left\{V\left(\hat{T}_{ht}\right)\right\}^2 - \Sigma_2 \; p(s_t)\left\{V\left(\hat{T}_{ht}\right)\right\}^2$$

$$+\Sigma_2 \; p(s_t)\left\{ \hat{V}_{ht} - V\left(\hat{T}_{ht}\right) \right\}^2$$

$$= E\left\{ \hat{V}_s - V\left(\hat{T}_{ht}\right) \right\}^2 + C \tag{3.3.8}$$

where,

$$C = \Sigma_2 \left[p(s_t) \left\{\hat{V}_{ht} - V\left(\hat{T}_{ht}\right)\right\}^2 - \Sigma_2 \; p(s_t)\left\{V\left(\hat{T}_{ht}\right)\right\}^2 \right]$$

$$= \Sigma_2 \; p(s_t)\left[\left\{ \hat{V}_{ht} - V(\hat{T}_{ht}) \right\}^2 - \left\{V(\hat{T}_{ht})\right\}^2\right] \tag{3.3.9}$$

Since Σ_2 is the summation over those s_t's for which $\hat{V}_{ht} < 0$, $C \geq 0$. Thus, the second moment of $\hat{V}_s$ about $V\left(\hat{T}_{ht}\right)$ is always less than or equal to the variance of $\hat{V}_{ht}$.

If this second moment is also less than the variance of the estimator of the variance due to Sen, Yates and Grundy, one may prefer to use Sen's modified estimator even when it is biased. On this criterian, one should use Sen's modified estimator to estimate the variance of the estimator $\hat{T}_{ht}$ of the total of the following population with p_i's as given in the table below.

Table 3.3.1

Unit	x	p
1	3.0	0.10
2	2.0	0.15
3	7.0	0.20
4	5.0	0.25
5	8.0	0.30
Total	25.0	1.00

This may be verified by the readers (cf. Sen, 1953, Tables I and II, pp. 124-125). The general discussion by Subrahmany (1970, Section 5, p. 112) throws the some more light, in addition to confirming these results, when the variance estimators are negative.

We now discuss the results of Seth (1966) and Lanke (1975).

Lemma 3.3.2

In addition to the notation already introduced in this chapter, we need the following. Given a smapling design p we define the exclusion probabilities $\pi_1^*, \ldots, \pi_N^*$ by

$$\pi_i^* = \sum_{s \not\ni i} p(s)$$

and analogously

$$\pi_{ij}^* = \sum_{s \not\ni i \cap s \not\ni j} p(s);$$

thus we have

$$\pi_i^* = 1 - \pi_i$$

$$\pi_{ij}^* = 1 - \pi_i - \pi_j + \pi_{ij}.$$

The easily checked relation

$$\pi_i^* \pi_j^* - \pi_{ij}^* = \pi_i \pi_j - \pi_{ij} \tag{3.3.10}$$

will be useful in sequel.

In the special case $n' = 2$, Seth (1966) gives the following theorem, which has obvious implications concerning non-negativity of $\hat{V}_{syg}$.

Theorem 3.3.1

Let p' be a sampling design of fixed size n' such that its first and second order inclusion probabilities π_i' and π_{ij}' satisfy

$$\pi_{ij}' \le \pi_i' \pi_j' \qquad \text{for all i and j}$$

Further, let p the sampling design which is performed as follows: first draw a sample s' according to p', then draw a sample s'' of size n'' with simple random sampling without replacement from $u - s'$ and finally put $s = s' \cup s''$. Then the inclusion probabilities of p satisfy.

$$\pi_{ij} \le \pi_i \pi_j \quad \text{for all i and j}$$

Proof :

It is easy to check that Seth's proof for the case $n' = 2$ in fact works for general n'. On the other hand, the general version given above is even simpler to prove than Seth's special case : in the general case it is sufficient to give a proof when $n''=1$ since, such a proof given, the result follows for general n'' by repeated use of the part already proved. A further simplication is obtained if we work with exclusion probabilities instead of inclusion probabilities. Then we immediately get

$$\pi_i^* = \pi_i'^* \frac{N-n'-1}{N-n'}$$

$$\pi_{ij}^* = \pi_{ij}'^* \frac{N-n'-2}{N-n'}$$

hence,

$$\pi_i^* \pi_j^* - \pi_{ij}^* = \left(\pi_i'^* \pi_j'^* - \pi_{ij}'^*\right)\left(\frac{N-n'-1}{N-n'}\right)^2 + \pi_{ij}'^* .$$

Now the relation (3.3.10) gives the result.

In order to calculate the estimator, its variance and the estimator of variance, we need the values of π_i and π_{ij} for a given scheme of sampling. By way of illustration, we obtain these values for certain well known sampling schemes.

3.3.4 Calculation of Probabilities of Inclusion for Certain Sampling Schemes

For Midzuno's scheme of sampling where the probability of i-th unit (i = 1, 2, . . ., N) being drawn at the first draw is p_i and at a subsequent draw is 1/(N–1) given that some other units have occurred at the previous draws, we have

$p_{i1} = p_i$ and for $r > 1$.

$$p_{ir} = (1-p_i) \mathop{\pi}_{s=2}^{r-1} \frac{N-s}{N-s+1} \cdot \frac{1}{N-r+1}$$

$$= \frac{(1-p_i)}{N-1};$$

from Eq. (2.8.17) of Chapter 2. Therefore, we get

$$\pi_i = \sum_{r=1}^{n} p_{ir} = p_i + \frac{n-1}{N-1}(1-p_i) \qquad (3.3.11)$$

Also,

π_{ij} = (Prob. of i-th unit or j-th unit occurring at the first draw and the remaining unit occurring at a subsequent draw) + (Prob. of i-th and j-th units occurring after the first draw).

$$=(p_i + p_j)\ \frac{n-1}{N-1} + (1-p_i\ p_j)\ \frac{(n-1)(n-2)}{(N-1)(N-2)} \qquad (3.3.12)$$

by using Eqs. (2.2.27) and (2.2.29).

It may be noted that Midzuno's scheme of sampling gives to a given sample the probability proportional to the total size of the units in the sample provided p_i is proportional to the size of the i-th unit for all i.

For Horvitz and Thompson's sampling scheme where n = 2.

$$\pi_i = p_{i1} + p_{i2}$$

$$= p_i \left(1 + S - \frac{p_i}{1-p_i}\right) \qquad (3.3.13)$$

From Eq. (2.2.45) of Chapter 2 with

$$S = \sum_{j=1}^{N} \frac{p_j}{1-p_j}$$

Also,

$$\pi_{ij} = p_i p_j \left(\frac{1}{1-p_i} + \frac{1}{1-\ p_j}\right) \qquad (3.3.14)$$

The expression in Eqs. (3.3.11) to (3.3.14) are due to Horvitz and Thompson (1952). The reader can now easily verify Sen's theorem, stated in Sub-section 3.3.2, to the effect that Sen, Yates and Grundy's estimator of variance is positive for the above two sampling schemes. That, this estimator of variance is the positive for n=N–1 for any sampling scheme (Seth, 1966), can also be easily seen. A reader may verify these results empirically with the help of data in Table 3.3.1.

For $n \geq 2$, the recurrence relations are given by Singh (1954, Eqs. on pp. 49-50) for a generalised sampling scheme, which reduces to Horvitz and Thompson's scheme where n = 2. But the expressions are complicated and therefore the computations for n > 2 and for moderately large N are difficult for such a sampling scheme.

Hartley and Rao (1962), while considering an alternate scheme of sampling to resolve the above difficulty, has commented on the work of Des Raj (1956) that his assumption, $x_i = \alpha + \beta y_i, y_i$ being an auxiliary variate, in finding those values of π_{ij} which minimise the variance of $\hat{T}_{ht}$ in Eq. (3.3.1); nullfy the utility of his results, because, in that case the regression estimator is the best estimator. The argument is not enough to justify the comment. For, from the works of Tikkiwal (1960) and Prabhu Ajgaonkar (1962), the regression estimator can be shown to be the best estimator in T_1-class where as the estimator considered by Des Raj is the best estimator in T_2-class. Therefore, a direct comparison of the variance of the two estimtors is necessary in order to examine the relative efficiency of the two estimators.

3.4 SAMPLING WITH VARYING PROBABILITIES AND WITH REPLACEMENT

We now present the results for varying probability sampling with replacement in general. Let $\hat{T}_{2w}$ denote the estimator for sampling with replacement.

In this case,

$$\pi_{ij} = \sum_{r \neq s}^{n} p(ir, js) = \pi_i \pi_j - \sum_{r=1}^{n} p_{ir} p_{jr}$$

Therefore, from Theorem 3.2.1

$$V(\hat{T}_{2w}) = \sum_{i=1}^{N} \frac{x_i^2}{\pi_i} + \sum_{i=1}^{N} \sum_{j=1}^{N} x_i x_j \left(1 - \frac{\sum_{r=1}^{n} p_{ir} p_{jr}}{\pi_i \pi_j}\right) - T^2$$

$$= \sum_{i=1}^{N} \frac{x_i^2}{\pi_i^2} (\pi_i - p_{ir}^2) - \sum_{i \neq j} \frac{x_i x_j}{\pi_i \pi_j} \left\{ \sum_{r=1}^{n} p_{ir} p_{jr} \right\} \tag{3.4.1}$$

In many practical situations, $p_{ir} = p_i$ for $r = 1, 2, \ldots, n$ and therefore $\pi_i = np_i$.

Substituting these values in (3.4.1)

$$V(\hat{T}_{2w}) = \frac{1}{n} \sum_{i=1}^{N} \left(\frac{x_i^2}{p_i} - T^2 \right) \tag{3.4.2}$$

In order to obtain an unbiased estimator of the variance of $\hat{T}_{2w}$, we note

$$E\left[\sum_{r=1}^{n}\left(\beta_{ro}^{st}\right)^2 X_r^2 - \frac{\hat{T}_{2w}^2}{n}\right]$$

$$= \sum_{i=1}^{N}\frac{x_i^2}{np_i} - \frac{1}{n}\left(\mathrm{Var}\left(\hat{T}_{2w}\right) + T^2\right)$$

$$= \left(\frac{n-1}{n}\right)\mathrm{Var}\left(\hat{T}_{2w}\right)$$

Therefore, an unbiased estimator of the variance of $\hat{T}_{2w}$ is

$$\frac{n}{n-1}\left(\sum_{r=1}^{n}\left(\beta_{ro}^{st}\right)^2 X_r^2\ \hat{T}_{2w}^2\right) \tag{3.4.3}$$

We now compare the efficiency of the general sampling scheme with replacement with the sampling scheme without replacement in Section 3. Under the assumption that the probability of inclusion, under the two sampling schemes, in the same. From Eqs. (3.3.1) and (3.4.1), we have

$$V\left(\hat{T}_{ht}\right) - V\left(\hat{T}_{2w}\right) = \left\{\sum_{i=1}^{N}\frac{x_i^2}{\pi_i^2}\sum_{r\neq s}^{n} p_{ir}\,p_{is} + \sum_{i\neq j}^{N}\frac{x_i\,x_j}{\pi_i\,\pi_j}\left(\sum_{r\neq s}^{n} p_{ir}\,p_{js} - \pi_{ij}\right)\right\} \tag{3.4.4}$$

a quadratic form which is negative definite when there are equal probabilities of selection in the two sampling systems and when in the population, x_i's are not all equal. This shows that sampling with replacement cannot be more efficient than sampling without replacement for equal probabilities of selection, a well known classical result. When there is varying probabilities of selection, no general statement can be made.

Durbin (1953, section 4, p. 266) states, without proof that the quadratic expresion in Eq. (3.4.1) is usually negative, but it is easy to invent cases in which it is positive. Folligi (1963) discusses some important sampling schemes, with $\pi_i = np_i$, where the quadratic expression is indeed negative. This justifies, to some extent, Durbin's statement. However, it needs further examination.

3.5 UNEQUAL PROBABILITY SAMPLING WITHOUT REPLACEMENT DUE TO RAO, HARTLEY AND COCHRAN

Before presenting the results due to Rao, Hartley and Cochran, it is instructive to deal with the following preliminaries.

(a) Certain Preliminaries

Let us first consider cluster sampling, where n clusters are selected at random, out of N clusters by simple random sampling without replacement scheme to estimate the population total T. Let each of the N clusters be of size M.

Let

$$x_{i.} = \frac{1}{M} \sum_{j=1}^{M} x_{ij} \quad \text{for} \quad 1 = 1, 2, \ldots, N.$$

$$x_{..} = \frac{1}{NM} \sum_{i=1}^{N} \sum_{j=1}^{M} x_{ij}$$

$$S_b^2 = \frac{1}{N-1} \Sigma \, (x_{i.} - x_{..})^2$$

$$S^2 = \frac{1}{NM-1} \Sigma\Sigma \, (x_{ij} - x_{..})^2$$

$$\hat{T} = \frac{N}{n} \sum_{i=1}^{n} x_i$$

Let $\hat{T}$ be the estimator of T, then

$$V(\hat{T}) = N^2 \frac{N-n}{Nn} S_b^2$$

If the clusters are formed randomly by selecting M units from the population with the help of simple random sampling without replacement scheme; then

$$V(\hat{T}) = N^2 \frac{N-n}{Nn} E\left(S_b^2\right)$$

$$= N^2 \frac{N-n}{Nn} \frac{1}{N-1} \sum_{i}^{N} E\,(x_{i.} - x_{..})^2$$

$$= N^2 \frac{N-n}{Nn} \frac{N}{N-1} \left(\frac{NM-M}{NM^2} S^2\right)$$

$$=N^2 \frac{NM-nM}{NM} \frac{S^2}{nM}$$

which is the same as if nM units are randomly selected out of NM units by SRSWOR scheme. Thus cluster sampling is as good as SRSWOR scheme, if the clusters are randomly formed.

Let us know consider stratified sampling, in which strata are randomly formed. Then the variance of $\hat{T}$, the estimator of T in this case of sampling is given by

$$V\left(\hat{T}\right) = \Sigma \frac{N_i - n_i}{N_i\, n_i} N_i^2\, E\left(S_i^2\right)$$

in the usual notations, when k, N_i's and n_i's are fixed in advance. Now

$$E\left\{(N_i - 1)\, S_i^2\right\} = E \sum_{j=1}^{N_i} (x_{ij} - x_{ir})^2$$

$$= E\left\{\sum_{j=1}^{N_i} (x_{ij} - x_{..})^2 - N_i\, (x_{ir} - x_{..})^2\right\}$$

$$= N_i \frac{N-1}{N} S^2 - N_i \frac{N-N_i}{NN_i} S^2$$

$$= \frac{N_i}{N} S^2 \left\{(N-1) - \frac{N-N_i}{N_i}\right\}$$

$$= \frac{N_i}{N} S^2 \left\{\frac{N(N_i-1)}{N_i}\right\} = (N_i - 1)\, S^2$$

And under proportional sampling with in each stratum i.e.,

$n_i \,\alpha\, N_i$

We have $n_i = N_i \frac{n}{N}$.

Therefore,

$$V\left(\hat{T}\right) = \sum_{i=1}^{k} \frac{N_i - \frac{n}{N} N_i}{N_i \,.\, N_i \frac{n}{N}} N_i^2\, S^2$$

$$= \sum_{i=1}^{k} \frac{N_i\left(1-\frac{n}{N}\right)}{n/N} S^2 = \frac{N-n}{n} S^2 \sum_{i=1}^{k} N_i = N^2 \cdot \frac{N-n}{Nn} S^2$$

which is the same as if n units are randomly selected out of N units by SRSWOR scheme. Thus, if the population is divided into k strata at random and if there is proportional sampling within each strata, then stratified sampling is as good as SRSWOR scheme.

(b) Rao-Hartley-Cochran Results

In the light of the above analysis, we consider the sampling procedure with varying probabilities and without replacement due to Rao, Hartley and Cochran (1962). In order to draw a sample of size n according to this sampling procedure, the population under study is divided at random into n strata, the i-th stratum being of the size N_i for i = 1, 2, . . ., n so that $\sum_{i=1}^{n} N_i = N$, the size of the population. Let x_{i_j} for j = 1, 2, . . ., N_i denote the value of i_j-th unit falling in the i-th stratum. One unit from each stratum is selected so as to have the ultimate sample of size n. Let the conditional probability of i_j-th unit being selected from the i-th stratum be p_{i_j}/π_i, where

$$\pi_i = \sum_{j=1}^{N_i} p_{i_j} \quad \text{and} \quad \sum_{i=1}^{n} \pi_i = 1$$

Then Eq. (3.2.2) the minimum variance linear unbiased estimator of T_i, the total of the i-th stratum is

$$\beta_{10}^{i} X_1^{i} \quad \text{where,} \quad \beta_{10}^{i} = \pi_i / p_{i_j} \quad \text{if} \quad X_1^{i} = x_{i_j}$$

Therefore, an unbiased estimator of T is given by

$$\hat{T}_{rhc} = \sum_{i=1}^{n} \beta_{10}^{i} X_1^{i} \tag{3.5.1}$$

Further,

$$V(\hat{T}_{rhc.}) = E\left\{\sum_{i=1}^{n}\left(\beta_{10}^{i} X_1^{i} - T_i\right)\right\}^2$$

$$= E_1 \sum_{i=1}^{n} E_2\left(\beta_{10}^{i} X_1^{i} - T_i\right)^2$$

where, E_2 is the conditional expectation for a given set of n strata and E_1 is the expectation over all random sets of n strata. Therefore,

$$V\left(\hat{T}_{rhc}\right) = E_1 \sum_{i=1}^{n} \sum_{j=1}^{N_i} \left(\frac{x_{i_j}\pi_i}{p_{i_j}} - T_i\right)^2 \frac{p_{i_j}}{\pi_i}$$

$$= E_1 \sum_{i=1}^{n} \left\{ \sum_{j=1}^{N_i} \sum_{j'(\neq j)}^{N_i} \frac{x_{i_j}^2}{p_{i_j}} p_{i_{j'}} - \sum_{j\neq j'}^{N_i} x_{i_j} x_{i_{j'}} \right\}$$

when the strata are randomly formed, the probability that i_j-th and $i_{j'}$-th units are included in the i-th stratum of size N_i is $N_i(N_i-1)/N(N-1)$. Therefore,

$$V\left(\hat{T}_{rhc}\right) = \sum_{i=1}^{n} \frac{N_i(N_i-1)}{N(N-1)} \left\{ \sum_{j=1}^{N} \frac{x_j^2}{p_j}(1-p_j) - \sum_{j\neq j'}^{N} x_j x_{j'} \right\}$$

$$= \frac{\sum_{i=1}^{n} N_i^2 - N}{N(N-1)} \left(\sum_{j=1}^{N} \frac{x_j^2}{p_j} - T^2 \right) \qquad (3.5.2)$$

a result due to Rao *et al* (1962, Eq. 5, p. 484). The variance of $\hat{T}_{rhc}$ is minimum when all N_i's are equal, giving $N_i = N/n$ for $i = 1, 2, \ldots, n$. Thus, N must be a multiple of n for variance of $\hat{T}_{rhc}$ to be minimum. When N is not a multiple of n, let $N = nR + k$, where $0 < k < n$ and R is positive integer. Then we may choose $N_i = R+1$, $i = 1, 2, \ldots, k$ and $N_i = R, i > k$, for random stratification. The variance of $\hat{T}_{rhc}$, so obtained, is shown by Rao *et al* to be close to the minimum. For k=1, n–1. In order to obtain an unbiased estimator of the variance of $\hat{T}_{rhc}$, we note

$$E\left\{ \sum_{i=1}^{n} \left(\beta_{10}\, x_1^i\right)^2 / \pi_i - \hat{T}_{rhc}^2 \right\}$$

$$= \sum_{i=1}^{n} E_1 \sum_{j=1}^{N_i} \left(x_{i_j}^2 / p_{i_j} \right) - V\left(\hat{T}_{rhc}\right) - T^2$$

$$= \left\{ \sum_{j=1}^{N} \left(x_j^2 / p_j \right) - T^2 \right\} - V\left(\hat{T}_{rhc}\right)$$

$$= \left\{ \frac{N(N-1)}{\Sigma N_i^2 - N} - 1 \right\} V\left(\hat{T}_{rhc}\right)$$

$$= \frac{N^2 - \Sigma N_i^2}{\Sigma N_i^2 - N} V\left(\hat{T}_{rhc}\right)$$

Therefore, an unbiased estimator of the variance of $\hat{T}_{rhc}$ is

$$\frac{\Sigma N_i^2 - N}{N^2 - \Sigma N_i^2} \left\{ \sum_{i=1}^{n} \left(\beta_{10} X_1^i\right)^2 / \pi_i - \hat{T}_{rhc}^2 \right\} \tag{3.5.3}$$

an expression similar to that given by Rao, Hartley and Cochran (1962, Eq. 15, p. 485), when N_i's are all equal, selection of one unit from each random stratum amounts to proportional allocation of n units to n strata. If the analogy with simple random sampling, is to hold good, in this case, the estimator $\hat{T}_{rhc}$ should be as efficient as random sampling with varying probabilities and without replacement. This is not true in the light of the work of Gupta (1964) showing that the estimator in the latter case is some times more efficient than the estimator due to Rao *et al.* However, it is always more efficient than sampling with varying probabilities and with replacement in view of the following relation.

$$V\left(\hat{T}_{rhc}\right) = \frac{N-n}{N-1} V\left(\hat{T}_2\right)_w \tag{3.5.4}$$

which is true in view of the Eq. (3.5.2) when N_i's are all equal.

3.6 THE THEORY OF MULTISTAGE SAMPLING WITH VARYING PROBABILITIES AND WITH OR WITHOUT REPLACEMENT

The Theory of multi-stage sampling with varying probabilities and with or without replacement has been discussed mainly by Hansen and Hurwitz (1943, 1949), Horvitz and Thompson (1952), Sukhatme and Narain (1952), Durbin (1953), Ecimovic (1956) and Tikkiwal (1965). Though there were many ways of extension. Tikkiwal defined a particular extended T_2-class of linear estimators for developing the theory of multi-stage sampling and showed that the estimators developed by the various earlier authors belongs to this class. Because of this elegence, we present his approach to the theory and then discuss the various important developments in this area as a corollary of his approach to the theory.

We first discuss the theory of two-stage sampling in detail and then indicate its generalisation to multi-stage sampling.

Let x_{ij} denote the value of j-th second-stage unit (ssu) in the i-th first-stage unit (fsu) for i = 1, 2, . . ., N and j = 1,2, . . ., M_i. Let us estimate the population total $T = \sum_{i=1}^{N} \sum_{j=1}^{M_i} x_{ij}$. For this, let us draw fsu and m_i ssu from the M_i ssu of the i-th fsu if it is selected at a particular first-stage draw, for i= 1, 2, . . ., N. Thus, in sampling with replacement; where a given fsu say i-th, may occur more than once, say t times, in the sample t independent samples of size m_i ssu will be drawn out of M_i ssu in the i-th fsu. Let X_{rs} denote the variate value of the ssu drawn, at the s-th second-stage draw, from the ssu's of the fsu selected at the r-th first stage draw for r = 1,2, . . ., n and s = 1,2, . . ., m_i if i-th first-stage unit occurs at the r-th for first stage draw for i = 1, 2, . . ., N. Let p_{js}^{ir} denote the probability of i-th first-stage unit occurring at the r-th first-stage draw and then j-th second-stage unit, in the i-th first-stage unit, occurring at the s-th second-stage draw. Let p_{ir} denote the probability of i-th first-stage unit occurring at the r-th first-stage draw. Let $p(js \mid ir)$ denote the conditional probability of j-th second stage unit, in the i-th first-stage unit, occurring at the s-th second-stage draw having occurred i-th first-stage unit at the r-th first-stage draw. Let $\hat{T}_2$, an estimator in T_2-class, of population total T, be given by

$$\hat{T}_{2,ms} = \Sigma\Sigma \beta_{so}^{r} x_{rs}, \tag{3.6.1}$$

where, $\beta_{so}^{r} = \beta_j^i$ if i-th first-stage unit is selected at the r-th first-stage draw and then j-th second-stage unit, in the first-stage unit, is selected at the s-th second-stage draw.

Now,

$$E\left(\hat{T}_{2,ms}\right) = \sum_{i=1}^{N} \sum_{j=1}^{M_i} \left(x_{ij}\, \beta_j^i\right) \left(\sum_{r=1}^{n} \sum_{s=1}^{m_i} p_{js}^{ir} \right)$$

$$= \sum_{i=1}^{N} \sum_{j=1}^{M_i} x_{ij}\, \beta_j^i\, \pi_j^i$$

where, $\pi_j^i = \sum_{r=1}^{n} \sum_{s=1}^{m_i} p_{js}^{ir}$.

If this is to be unbiased estimator of the population total T,

$$\beta_j^i = \frac{1}{\pi_j^i}$$

a unique value for β_j^i. Thus, there is only one unbiased estimator in T_2-class and therefore it is minimum variance linear unbiased estimator. The variance and its estimator, for the minimum variance-linear unbiased estimator in Eq. (3.6.1) for sampling with or without replacement or both at the two stages of sampling, can be obtained as in Section 3.1.

In order to discuss the various results obtained by Durbin (1953) and others on multi-stage sampling, let p (js | ir) be independent of r. This assumption would mean that the sampling system adopted at the second-stage is same for all the first-stage draws, which is reasonable.

Under the above practical assumption,

$$\pi_j^i = \sum_{r=1}^{n} \sum_{s=1}^{m_i} p_{ir} \; p(js \mid ir)$$

$$= \pi_i \sum_{s=1}^{m_i} p(js \mid ir)$$

Therefore, the estimator in Eq. (3.6.1) is equivalent to

$$\hat{T}_{2,ms} = \sum_{r=1}^{n} \beta_{ro}^{st} \; X_r \tag{3.6.2}$$

where, $\beta_{ro}^{st} = 1/\pi_i$, and where $X_r = \hat{T}_i$ the estimator of the total T_i of the i-th first-stage unit; when i-th first-stage unit occurs at the r-th first-stage draw. If there are more than two stages, that the estimator can be put in the form in Eq. (3.6.2) is now obvious. In this form the result was given for the first time by Durbin (1953). Thus under the above practical assumption, Durbin's estimator lies in T_2-class.

The estimator given by Sukhatme and Narain (1952) for sampling with replacement in their case (ii); where m_i second-stage units are selected each time, with simple random sampling without replacement, from the total M_i second-stage units in the i-th first-stage unit, whenever it is selected in a first-stage draws; is a special case of Durbin's form and therefore, it also belongs to T_2-class. As regards the other estimator given by these authors for their case (i) where, if a given first-stage unit, say i-th, is selected t times in n draws, then a sample of tm_i second-stage units are drawn out of M_i second-stage units with simple random sampling without replacement; it is easily seen that this estimator is not of the forms given in Eqs. (3.6.1) and (3.6.2) and so it does not belong to this T_2-class (Author, 1984, section 2).

The above two estimators for two different types of sampling have been discussed in detail by the first author, Sukhatme (1953, section 8.2, p. 358 and section 8.9, p. 379. See also the revised editions 1970 and 1984).

The various unbiased estimators considered by Ecimovic (1956) are obtained by him from the unbiased estimator, of the total, given in his Eq. (1), when there is simple random sampling without replacement at all the three stages. Since this unbiased estimator is also special case of Durbin's form, the various estimator due to Ecimovic belong to T_2-class.

Hansen and Hurwitz (1943, p. 341) and later Horvitz and Thompson (1952, pp. 673-675) have considered a self-weighting design where m_i second stage units are so chosen that the weight associated to the unit x_{ij} is constant for all i, j. These estimators for such a design are easily obtained by taking $\pi_j^i = t$, a constant quantity. Thus the estimators considered by these authors also belong to T_2-class.

Theoretically, it is possible to have sampling systems where $p(js \mid ir)$ is not independent of r. In such a case, the estimators in Eqs.(3.6.1) and (3.6.2) are not identical. In order to choose between the two estimators, one has to compare the efficiencies of the two estimators for the sampling system adopted. Since such sampling systems are not in use, we confine our attention to the case where $p(js \mid ir)$ is independent of r where the estimators in (3.6.1) and (3.6.2) are identical. We proceed to find the variance of $\hat{T}_{2,ms}$ in (3.6.2) and an estimator of its variance, when there is sampling with or without replacement. Let

$$E\left[\hat{T}_{2,ms} \mid r = 1, 2, \ldots, n\right] = \hat{T}'_{2,ms}$$

an estimator of T when there is only one stage of sampling. Then, for sampling with or without replacement;

$$V\left(\hat{T}_{2,ms}\right) = E\left(\hat{T}_{2,ms} - \hat{T}'_{2,ms} + \hat{T}'_{2,ms} - T\right)^2$$

$$= \text{Var}\left(\hat{T}'_{2,ms}\right) + E\left(\hat{T}_{2,ms} - \hat{T}'_{2,ms}\right)^2$$

$$= V\left(\hat{T}'_{2,ms}\right) + E_1 E_2 \left\{ \sum_{r=1}^{n} \beta_{ro} X_r - E\left(\sum_{r=1}^{n} \beta_{ro} X_r \mid r = 1, 2, \ldots, n \right) \right\}^2$$

$$= V\left(\hat{T}'_{2,ms}\right) + E_1 \left[\sum_{r=1}^{n} E_2 \{\beta_{ro} (X_r - E(X_r) \mid r = 1, 2, \ldots, n)\}^2 \mid (r = 1, 2, \ldots, n) \right]$$

$$= V\left(\hat{T}'_{2,ms}\right) + E_1 \left\{ \sum_{r=1}^{n} \sum_{i=1}^{N} \frac{V(\hat{T}_i \mid i)}{\pi_i^2} p_{i_r} \right\}$$

$$= V\left(\hat{T}'_{2,ms}\right) + \sum_{i=1}^{N} \frac{V(\hat{T}_i \mid i)}{\pi_i} \tag{3.6.3}$$

The estimation of the variance, in more than one stage of sampling, for sampling with or without replacement does not present any new problem other than already considered in Sections 3 and 4.

In order to see that this is so, let $\hat{V}_{wo}$ denote any one the two unbiased estimators, given in Section 3, of the variance in case of sampling without replacement. Let $\hat{V}_w$ denote some unbiased estimator of the variance, in case of sampling with replacement, such as one given in Section 4. Let $\hat{V}_i$ be an unbiased estimator of $V(\hat{T}_i \mid i)$. Let $\sum_{i \in s_1}$ denote the some over those first-stage units which occur in the sample. Let $x_i = \hat{T}_i$ in $\hat{V}_{wo}$ and $\hat{V}_w$. Then, for sampling without replacement, we have

$$E\left\{ \hat{V}_{wo} + \sum_{i \in s_1} (V_i / \pi_i) \right\}$$

$$= E\left(\hat{V}_{wo}\right) + \sum_{r=1}^{n} \sum_{i=1}^{N} \frac{V(\hat{T}_i \mid i)}{\pi_i} p_{ir}$$

$$= E\left(\hat{V}_{wo}\right) + \sum_{i=1}^{N} V(\hat{T}_i \mid i)$$

We now proceed to find $E\left(\hat{V}_{wo}\right)$. Since $E\left(\hat{T}_i^2 \mid i\right) = \text{Var}\left(\hat{T}_i \mid i\right) + T_i^2$ and $E\left(\hat{T}_i \hat{T}_j \mid i, j\right) = T_i T_j$ and since $\hat{V}_{wo}$ with x_i's as such is an unbiased estimator of $V\left(\hat{T}_{ht}\right)$ as given by Eq. (3.3.1) or by Eq. (3.3.3) we note

$$E\left(\hat{V}_{wo}\right) = \sum_{i=1}^{N} T_i^2 (1-\pi_i)/\pi_i + \sum_{i \neq j}^{N} \frac{T_i T_j}{\pi_i \pi_j} (\pi_{ij} - \pi_i \pi_j)$$

$$+ \sum_{i=1}^{N} V(\hat{T}_i \mid i)\,(1-\pi_i)/\pi_i$$

$$= V\left(\hat{T}'_{2,ms}\right) + \sum_{i=1}^{N} \frac{V(\hat{T}_i \mid i)}{\pi_i} - \sum_{i=1}^{N} V(\hat{T}_i \mid i)$$

Therefore,

$$E\left\{\hat{V}_{wo} + \sum_{i \in s_1} (\hat{V}_i / \pi_i)\right\} = V(\hat{T}'_{2,ms}) + \sum_{i=1}^{N} \frac{V(\hat{T} \mid i)}{\pi_i} = V(\hat{T}_{2,ms})$$

Thus, for sampling without replacement, the unbiased estimator of $V(\hat{T}_{2,ms})$.

$$\hat{V}_{wo} + \sum_{i \in s_t} (\hat{V}_i / \pi_i) \tag{3.6.4}$$

It may be noted that Theorem 6.3 (Des Raj, 1968, section 6.4, p. 118) follows by taking $V(\hat{T}_{2,ms})$, in Eq. (3.6.3) from Eq. (3.3.2) and by taking $\hat{V}_{wo}$, in Eq. (3.6.4) from Eq. (3.3.3). For sampling with replacement, we have from Eqs. (3.4.2) and (3.6.3).

$$E\left(\hat{V}_{w}\right) = \frac{1}{n}\left(\sum_{i=1}^{N} \frac{T_i^2}{p_i} - T^2\right) + \frac{1}{n}\sum_{i=1}^{N} \frac{\mathrm{Var}\,(\hat{T}_i \mid i)}{p_i}$$

$$= V\left(\hat{T}'_{2,ms}\right) + \sum_{i=1}^{N} \frac{V(\hat{T}_i \mid i)}{\pi_i}$$

$$= V\left(\hat{T}_{2,ms}\right)$$

as $np_i = \pi_i$ in this case. Therefore, an unbiased estimator of $V\left(\hat{T}_{2,ms}\right)$ in case of sampling with replacement is

$$\hat{V}_{w} \tag{3.6.5}$$

The reader may verify, as an exercise, that the variance of the estimator $\bar{y}_{s_2}$ (Sukhatme and Sukhatme, 1970, section 7, p. 286) for two-stage sampling with equal probabilities and without replacement at both stages and its estimator can be obtained from Eqs. (3.6.3) and (3.6.5). Similarly, we can obtain the results in sampling with varying probabilities and with replacement as given by Sukhatme and Sukhatme [sections (8.2) and (8.3), pp. 315-318].

Yates (1949, section 7.17, pp. 226-227) has given a rule for estimation of the variance in multi-stage sampling when there is simple random sampling without replacement atleast at the first-stage. The following generalised rule, due to Durbin (1953, p. 264), for sampling with varying probabilities and without replacement atleast at the first-stage is obvious from (3.6.4); if we take, in the rule, the contribution from each of the first-stage unit, in the sample, as the unbiased estimate of the variance of $(\hat{T}_i/\pi_i)$ given that i-th unit has occurred in the sample.

> "The estimate of variance in multi-stage sampling is the sum of two-parts. The first part is equal to the estimate of variance calculated on the assumption that the first-stage values have been measured without error. The second part is calculated as if the first-stage units selected were fixed strata, the contribution from each first-stage units being multiplied by the probability of that unit's inclusion in the sample."

Yate's rule follows as a special case of Durbin's rule if we take $\pi_i = n/N$ for $i = 1, 2, \ldots, N$.

We may note that the coefficient $1/\pi_i$ of $\hat{V}_i$ in the second part of the Eq. (3.6.4) is the coefficient occurring in the estimator of population total in Eq. (3.6.2). Thus, Eq. (3.6.4) can in fact be written as

$$\hat{V}_{wo} + \sum_{i \in s_t} \beta_{ro}^{s_t} \hat{V}_i$$

Thus, we may re-state the Durbin's rule in a slightly more general form (Des Raj, 1966, p. 392).

Get an unbiased estimator of the variance in single-stage sampling. Obtain a copy of it by substituting for x_i, its estimator $\hat{T}_i$. Also get the copy of the estimator of the stratum total in single-stage sampling by substituting $\hat{V}_i$ for x_i. The sum of the two copies is an unbiased estimator of the variance in the multi-stage case.

The reader should verify that the estimators of variances in case of two-stage sampling with equal probability given by the author

(1985) in the basic course follow from the above rule. Regarding the wider applicability of the above rule, the reader should refer to the original paper of Des Raj or his Text Book (1968).

3.7 COMPARISON OF THE EFFICIENCY OF ESTIMATORS FOR SAMPLING WITH VARYING PROBABILITIES WITHOUT REPLACEMENT AND WITH REPLACEMENT IN CASE OF MULTISTAGE DESIGNS

This comparison has been presented in Section 3.4, when there is single stage of sampling. It is mentioned therein that for those sampling schemes without replacement where $\pi_i = np_i$; the estimators for sampling without replacement are generally more efficient than the estimators used for sampling with replacement. For such sampling schemes adopted at the first-stage of selection, it is seen from Eq.(3.6.3) that the above statement is true for multi-stage designs also provided we use the same sampling scheme at the subsequent stages of selection or those sampling schemes in which

$$\sum_{i=1}^{n} \frac{V(\hat{T}_i \mid i)}{\pi_i} \quad \text{for SWOR}$$

$$\leq \sum_{i=1}^{n} \frac{V(\hat{T}_i \mid i)}{\pi_i} \quad \text{for SWR} \tag{3.7.1}$$

We now present a result due to Des Raj (1964) {Text book, 1968, section 6.6.1, p. 122} which enables to estimate the gain in efficiency due to sampling without replacement over sampling with replacement where $\pi_i = np_i$. From Eqs. (3.6.3) and (3.6.4), the estimator of the variance in case of sampling without replacement is

$$\sum_{i,j \in s_1} \left(\frac{\pi_i \pi_j - \pi_{ij}}{\pi_{ij}} \right) \left(\frac{\hat{T}_i}{\pi_i} - \frac{\hat{T}_j}{\pi_j} \right)^2 + \sum_{i \in s_1} \left(\frac{\hat{V}_i}{\pi_i} \right) \tag{3.7.2}$$

From Eqs. (3.4.1) and (3.6.3), the variance of $\hat{T}_{2,ms}$ in case of sampling with replacement is given by

$$V(\hat{T}_{2,ms}) = \frac{1}{n} \left(\sum_{i=1}^{N} \frac{T_i^2}{p_i} - T^2 \right) + \sum_{i=1}^{N} \frac{V(\hat{T}_i \mid i)}{\pi_i} \tag{3.7.3}$$

Now,

$$\sum_{i=1}^{N}\sum_{j>i}^{N}\left(\frac{T_i}{\pi_i}-\frac{T_j}{\pi_j}\right)^2 \pi_i \pi_j$$

$$= \sum_{i=1}^{N}\frac{T_i^2}{\pi_i}\sum_{j\neq i}^{N}\pi_j - 2\sum_{i=1}^{N}\sum_{j>i}^{N} T_i T_j$$

$$= \sum_{i=1}^{N}\frac{T_i^2}{\pi_i}(n-\pi_i) - 2\sum_{i=1}^{N}\sum_{j>i}^{N} T_i T_j$$

$$= n\sum_{i=1}^{N}\frac{T_i^2}{\pi_i} - \left(\sum_{i=1}^{N} T_i\right)^2$$

$$= n\left[\sum_{i=1}^{N}\frac{T_i^2}{\pi_i} - \frac{T^2}{n}\right] \tag{3.7.4}$$

Therefore,

$$V\left(\hat{T}_{2,ms}\right) = \frac{1}{n}\left(\sum_{i=1}^{N}\sum_{j>1}^{N}\pi_i \pi_j\left(\frac{T_i}{\pi_i}-\frac{T_j}{\pi_j}\right)^2\right) + \sum_{i=1}^{N}\frac{V(\hat{T}_i\,|\,i)}{\pi_i} \tag{3.7.5}$$

Suppose that we resort to sampling without replacement and then want to estimate the said gain, when sampling at subsequent stages is carried out in the same manner in both the sampling schemes. In other words the two sampling designs differ only at the first-stage. We already have the estimate of the variance for sampling without replacement. Therefore, we proceed to estimate the variance for sampling with replacement. For this

$$E\left[\sum_{i,j\in s_t}\frac{\pi_i \pi_j}{\pi_{ij}}\left(\frac{\hat{T}_i}{\pi_i}-\frac{\hat{T}_j}{\pi_j}\right)^2\right]$$

$$= E\left[\sum_{i,j\in s_t}\frac{\pi_i \pi_j}{\pi_{ij}} E\left\{\left(\frac{\hat{T}_i}{\pi_i}-\frac{\hat{T}_j}{\pi_j}\right)^2 |\, i,j\right\}\right]$$

$$= E\sum_{i,j\in s_t}\frac{\pi_i \pi_j}{\pi_{ij}}\left\{\left(\frac{T_i}{\pi_i}-\frac{T_j}{\pi_j}\right)^2 + \frac{V(\hat{T}_i\,|\,i)}{\pi_i^2} + \frac{V(\hat{T}_j\,|\,j)}{\pi_j^2}\right.$$

$$= \sum_{i=1}^{N} \sum_{j>i}^{N} \pi_i \pi_j \left(\frac{T_i}{\pi_i} - \frac{T_j}{\pi_j} \right)^2 + \sum_{i=1}^{N} \sum_{j>i}^{N} \pi_i \pi_j \left\{ \frac{V(\hat{T}_i \mid i)}{\pi_i^2} + \frac{V(\hat{T}_j \mid j)}{\pi_j^2} \right\}$$

$$= \sum_{i=1}^{N} \sum_{j>1}^{N} \pi_i \pi_j \left(\frac{T_i}{\pi_i} - \frac{T_j}{\pi_j} \right)^2 + \sum_{i=1}^{N} \frac{V(\hat{T}_i \mid i)}{\pi_i} \sum_{j(\neq i)} \pi_j$$

$$= \sum_{i=1}^{N} \sum_{j>i}^{N} \pi_i \pi_j \left(\frac{T_i}{\pi_i} - \frac{T_j}{\pi_j} \right)^2 + \sum_{i=1}^{N} \frac{n - \pi_i}{\pi_i} V(\hat{T}_i \mid i) \tag{3.7.6}$$

Thus, $\hat{V}_{wr}$ the estimator of the variance in case of sampling with replacement is given by

$$\hat{V}_{wr} = \frac{1}{n} \left\{ \sum_{i,j \in s_t} \frac{\pi_i \pi_j}{\pi_{ij}} \left(\frac{\hat{T}_i}{\pi_i} - \frac{\hat{T}_j}{\pi_j} \right)^2 - \sum_{i \in s_t} \left(\frac{n - \pi_i}{\pi_i^2} \right) \hat{V}_i + \sum_{i \in s_t} \frac{n \hat{V}_i}{\pi_i^2} \right\}$$

$$= \frac{1}{n} \left\{ \hat{V}_{wor} + \sum_{i,j \in s_t} \left(\frac{\hat{T}_i}{\pi_i} - \frac{\hat{T}_j}{\pi_j} \right)^2 \right\} \tag{3.7.7}$$

where, $\hat{V}_{wor}$ is the estimator of the variance in case of sampling without replacement was given by Eq. (3.7.2). We can now estimate the said gain from Eqs. (3.7.2) and (3.3.7). The reader may now attempt to estimate the said gain when we resort to sampling with replacement instead of sampling without replacement.

EXERCISES

3.1. Prove that the Horvitz-Thompson estimator $\hat{T}_{ht} = \sum_{i \in s} x_i / \pi_i$ is minimum variance linear unbiased estimator of population total in T_2-class.

3.2. (a) Show that the variance of the Horvitz-Thompson estimator, under a sampling plan where the probability of selection of unit i is π_i and the joint selection probability of unit i and j is π_{ij}, is

$$V\left(\hat{T}_{ht}\right) = \sum_{i=1}^{N} \sum_{j=1}^{N} (\pi_{ij} - \pi_i \pi_j) \frac{x_i}{\pi_i} \frac{x_j}{\pi_j}$$

(b) When $\pi_{ij} > 0$ for all pairs of units and the total sample size n is fixed, show that the following two estimators of this variance are design unbiased:

$$\hat{V}_{ht} = \sum_{i \in s} \sum_{j \in s} \frac{\pi_{ij} - \pi_i \pi_j}{\pi_{ij}} \frac{x_i}{\pi_i} \frac{x_j}{\pi_j}$$

and,

$$\hat{V}_{yg} = -\frac{1}{2} \sum_{i \in s} \sum_{j \in s} \frac{\pi_{ij} - \pi_i \pi_j}{\pi_{ij}} \left(\frac{x_i}{\pi_i} - \frac{x_j}{\pi_j}\right)^2$$

3.3. Specialize the general results of the Horvitz-Thompson estimator to show that with an SRSWOR scheme, a design unbiased estimator of the variance of $\hat{T} = N\bar{y}$ is $N^2(1 - n/N)s^2/n$, where $s^2 = \sum_{i \in s} (x_i - \bar{x})^2/(n-1)$.

3.4. (a) Under the SRSWOR scheme, show that

$$\pi_i = n/N \quad ; \quad i = 1, \ldots, N$$

and $$\pi_{ij} = \frac{n(n-1)}{N(n-1)} \quad ; \quad i \neq j = 1, \ldots, N$$

(b) Under the SRSWR, show that

$$\pi_i = 1 - \left(1 - \frac{1}{N}\right)^n \quad ; \quad i = 1, \ldots, N$$

and $$\pi_{ij} = 1 - 2\left(1 - \frac{1}{N}\right)^2 + \left(1 - \frac{2}{N}\right)^n \quad ; \quad i \neq j = 1, \ldots, N$$

3.5 (a) Let π_i^* $(i=1,2,\ldots,N)$ be the probability of not including the i-th unit in the sample and let π_{ij}^* $(i \neq j=1,2,\ldots,N)$ be the probability of including neither the i-th nor j-th units in the sample. Then show that

$$\pi_i^* \, \pi_j^* - \pi_{ij}^* = \pi_i \, \pi_j - \pi_{ij}$$

(b) Using (a) or otherwise prove the following:

Suppose p′ is a sampling design of fixed size n' such that its first and second order inclusion probabilities π_i' and π_{ij}' satisfy $\pi_{ij}' \leq \pi_j' \, \pi_j'$ for $i \neq j$. Further, let p be the sampling design where we first draw a sample s′ from the population u according to p′ and then draw a sample s″ of size n″ with simple random sampling without replacement from $u - s'$ and consider $s = s' \cup s''$ as the resulting sample of size $n = n' + n''$. Then the inclusion probability of p satisfy $\pi_{ij} \leq \pi_i \, \pi_j$ for all $i \neq j'$.

[Seth (1996), Lanke (1974)]

3.6. Consider a πps scheme in which the Horvitz-Thompson estimator $\hat{T}_{ht} = \sum_{i=1}^{n} x_i / \pi_i$ is used and the Hansen-Hurwitz strategy using the same set of $\{n, p_i\}$ values. Suppose that $\hat{V}_{pps}$, the customary variance estimator of $\hat{T}_{pps}$ is used to estimate $V(\hat{T}_{ht})$. Show that this is a biased estimator and that the bias is

$$n(n-1)^{-1}\left[V(\hat{T}_{pps}) - V(\hat{T}_{ht})\right].$$

This shows that when we use $\hat{V}_{pps}$ to estimate $V(\hat{T}_{ht})$, we tend to over estimate the variance whenever $V(\hat{T}_{ht})$ is smaller than $V(\hat{T}_{pps})$. Thus $\hat{V}_{pps}$ is a conservative estimator of $V(\hat{T}_{ht})$.

[Durbin (1953)]

3.7. Explain Midzuno scheme of sampling and show that under this scheme Sen, Yates and Grundy's estimate of variance is positive.

3.8. Let in a sampling with varying probabilities with replacement, a given sample s_t consists of $(i_1, i_2, \ldots, i_k)$, k-distinct units with i_j-th unit occurring R_{i_j} times. Show that the following estimator of population total T

$$\hat{T}_{2\,wr} = \frac{1}{n} \sum_{j=1}^{k} \frac{x_{i_j} \quad R_{i_j}}{p_{i_j}}$$

is MVLUE in T_2-class. Further, obtain combined unordered estimator $\hat{T}_{cu}$ corresponding to unordered estimator $\hat{T}_{2\,wr}$, and also obtain the expression of variance and unbiased estimate of the variance of $\hat{T}_{cu}$. For SRSWR, what will be the form of $\hat{T}_{cu}$?

[Bhargava (1978), Tikkiwal (1965), Basu (1958)]

3.9. A sample of n fsu's is drawn by SRSWOR from a population of N fsu's (the i-th fsu consisting of M_i ssu's) and from the i-th selected fsu m_i ssu's are selected by SRSWOR. Show that

$\bar{\bar{Z}} = \frac{1}{n} \sum_{i=1}^{n} \bar{z}_i$ is an unbiased estimator of population mean $\bar{x}_N$,

and it belongs to T_2-class.

where, $\bar{z}_i$ represent the arithmetic mean of the m_i selected ssu's in the i-th fsu.

Also obtain expressions of variance of $\bar{\bar{Z}}$ and estimate of the variance, using expression given in Eqs. (3.6.3) and (3.6.5)

3.10. A sample of n fsu's is drawn by varying probabilities with replacement from a population of N fsu's (the i-th fsu consisting of M_i ssu's) and from the i-th selected psu m_i ssu's are selected by SRSWOR; keeping in view that if i-th unit happens to be selected t_i times in a sample of n fsu's, t_i independent sub-samples of m_i ssu's each is drawn by SRSWOR. Show that

(a) $\bar{\bar{Z}} = \frac{1}{n} \sum_{i=1}^{n} \bar{z}_i$ (1)

is an unbiased estimator of population mean $\bar{x}_N$ and it belongs to T_2-class; where,

$$\bar{z}_i = \frac{1}{m_i} \sum_{j=1}^{m_i} z_{ij} \quad , \quad z_{ij} = \frac{M_i\, x_{ij}}{N\, \bar{m}\; p_i}$$

x_{ij} denote the value of the characteristic under study for the j-th ssu of the i-th fsu, $j = 1, \ldots, M_i$ and $i = 1, \ldots, N$.

p_i denote the selection probability assigned to i-th fsu of the population s.t. $\sum_{i=1}^{N} p_i = 1$.

(b) Obtain expressions of $V(\bar{\bar{z}})$ and an unbiased estimate of $V(\bar{\bar{z}})$, using expressions given in Eqs. (3.6.3) and (3.6.5).

3.11. Suppose n fsu's are selected with PPSWR and if the i-th unit occurs t_i times in the sample, then any of the following three procedures may be adopted in sub-sampling:

(i) $t_i\, m_i$ ssu's are selected with SRSWOR;

(ii) t_i independent samples of m_i units each, selected with SRSWOR, are drawn; and

(iii) m_i units are selected with SRSWOR and the observations are weighted by t_i.

Obtain the variances V_1, V_2 and V_3 of the estimators of the population mean based on the three procedures respectively and then show that for the same expected sample size

$$V_1 \le V_2 \le V_3$$

Also show that the estimators of the population means in the three procedures beong to T_2-class of linear estimators.

[Rao (1962)]

CHAPTER 4

Estimation in T_3 - Class

4.1 INTRODUCTION

We have defined, in part (a) of Chapter 1, a sample as an ordered set. Thus, a sampling mechanism, to obtain a sample, is one of unit by unit draw method. We have assumed all along, from Chapter 2 onward, that there is more than one draw, i.e., the sample size $n \geq 2$. For n=1; the sampling mechanism would result in those designs which are known as uni-cluster designs (Lanke, 1975, p. 12). Thus, from our study we have excluded such designs altogether. We are justified for doing so because of what follows.

By definition we regard two samples as different from each other if there is at least one unit in one of the two samples, which is not contained in the other. A sample design is said to be a uni-cluster design provided any two samples say s_1 and s_2 with $p(s_1) > 0$ and $p(s_2) > 0$ have no unit common between them. Without using the terminology of uni-cluster design, Godambe (1965, Theorem 9.1, p. 250) improved upon his earlier result (1955) regarding non-existence of MVLUE in his general class by stating that his theorem is true for designs other than the uni-cluster designs. He further stated that his theorem is, in fact, true for the class of even non-linear but unbiased estimators.

In view of Hanurao (1962), the uni-cluster designs would be generated by those sampling mechanisms in which there is only one draw. A good example of this is systematic sampling. Thus, we need to consider n = 1 in our treatment to this case. The only linear unbiased estimator in this case is $X \big| p(x)$ where X is the variate value of the unit occurring at the draw and $P(X)$ is the probability of that unit. Its variance is obtained by Eq. (3.2.3) by taking $\pi_i = P(x_i)$ and $\pi_{ij} = 0$. We now discuss linear estimation in T_3 – class.

Let there be a sample of size n drawn with a specific scheme of varying probabilities and with or without replacement out of N units in the population to estimate the population total $T = \sum_{i=1}^{N} x_i$. Let X_r

denote the variate value of the unit drawn at the r-th draw for r = 1,2 ,. . . , n. Then, from what is said in Chapter 1, the estimator in T_3 -class can be taken as

$$\hat{T}_3 = \beta^{s_t} \left(\sum_{r=1}^{n} X_r \right)_{s_t} \tag{4.1.1}$$

where β^{s_t} is the weight associated with the total of the units selected in the s_t -th sample for t = 1, 2,. . ., $n!\binom{N}{n}$ for sampling without replacement and t = 1, 2, . . . , N^n for sampling with replacement.

It is shown by Prabhu Ajgaonkar and Tikkiwal (1961) and Prabhu Ajgaonkar (1962) that there do not exist any minimum variance linear unbiased estimator in T_3 -class which is independent of population values for any sampling scheme including simple random sampling with or without replacement. We now prove below this result, by first presenting a Lemma, to be frequently used in this chapter.

4.2 A LEMMA IN QUADRATIC EQUATION

Lemma 4.2.1

Let $X'BX$ be a quadratic form with $X = (x_1, x_2, \ldots, x_N)'$ and

$$B = \begin{bmatrix} b_{11} & b_{12} & \cdots & b_{1N} \\ \cdot & & & \\ \cdot & & & \\ \cdot & & & \\ b_{N1} & b_{N2} & \cdots & b_{NN} \end{bmatrix}$$

where, $X \in R^{(N)}$ and $b_{ij} = b_{ji}$ for all i,j. Then $X'BX = 0$ for all $X \in R^{(N)}$, if and only if $B \equiv \phi$ (a null matrix).

Proof : Let

$$X'BX = (x_1, x_2, \ldots, x_N) \begin{bmatrix} b_{11} & b_{12} & \cdots & b_{1N} \\ \cdot & \cdot & \cdot & \\ \cdot & \cdot & & \\ \cdot & \cdot & & \\ b_{N1} & b_{N2} & \cdots & b_{NN} \end{bmatrix} \begin{bmatrix} x_1 \\ x_2 \\ \cdot \\ \cdot \\ \cdot \\ x_N \end{bmatrix} = 0$$

or,

$$\sum_{p=1}^{N} \sum_{q=1}^{N} b_{pq}\, x_p\, x_q = 0$$

or,

$$b_{11} x_1^2 + 2b_{12} x_1 x_2 + \ldots + 2b_{1N-1} x_1 x_{N-1} + 2b_{1N} x_1 x_N$$
$$+ b_{22} x_2^2 + 2b_{23} x_2 x_3 + \ldots + 2b_{2N} x_2 x_n$$
$$+ b_{N-1,N-1} x_{N-1}^2 + 2b_{N-1,N} x_{N-1} x_N + b_{N,N} x_N^2 = 0 \qquad (4.2.1)$$

If it is true for all $X \in R^{(N)}$, let

$$X = \begin{bmatrix} 1 \\ 0 \\ 0 \\ . \\ . \\ 0 \end{bmatrix} \qquad (4.2.2)$$

Then substituting this value of vector X in Eq. (4.2.1), we get

$$b_{11} = 0$$

Similarly to prove $b_{jj} = 0$ for any other j, we choose vector X with $x_j = 1$ and $x_i = 0$ for all $i \neq j$. Thus, $b_{jj} = 0$, j = 1, 2, . . ., N. Now if we take

$$X = \begin{bmatrix} 1 \\ 1 \\ 0 \\ . \\ . \\ 0 \end{bmatrix}$$

then,

$$b_{12} = 0$$

Similarly we can prove $b_{ij} = 0$ for any pair of (i, j) for $i \neq j$ by considering the vector X with $x_i = x_j = 1$ and $x_k = 0$ for $k \neq i, j$. This proves B = ϕ which, in turn, proves the necessary part of the Lemma. Also when B = ϕ (a null matrix) then $X'BX = 0$, whatever X may be. This proves the sufficient part of the Lemma.

We now turn to our discussion on MVLUE in T_3-class of estimation for sampling without replacement in Sections 4.3 to 4.5.

4.3 NON-EXISTENCE OF MVLUE IN THE CLASS FOR SAMPLING WITHOUT REPLACEMENT

We state and prove the following theorem.

Theorem 4.3.1

The minimum variance linear unbiased estimator, independent of population values, does not exist in T_3-class for sampling without replacement (SWOR).

(i) **Proof of a special case** We first prove, for simplicity, the theorem for a specified population (x_1, x_2, x_3, x_4), when a sample of size 2 is selected from the population by a sampling scheme without replacement. The possible samples are

$$\begin{aligned} s_1 &= (x_1, x_2); & s_2 &= (x_1, x_3); & s_3 &= (x_1, x_4) \\ s_4 &= (x_2, x_1); & s_5 &= (x_2, x_3); & s_6 &= (x_2, x_4) \\ s_7 &= (x_3, x_1); & s_8 &= (x_3, x_2); & s_9 &= (x_3, x_4) \\ s_{10} &= (x_4, x_1); & s_{11} &= (x_4, x_2); & s_{12} &= (x_4, x_3) \end{aligned}$$

For $s_t = (x_{i_1}, x_{i_2})$, let $\beta^{s_t} = W_{i_1 i_2}$. Then

$$\hat{T}_3 = W_{i_1 i_2} (x_{i_1} + x_{i_2}) \tag{4.3.1}$$

and,

$$E(\hat{T}_3) = \sum_{t=1}^{12} W_{i_1 i_2} (x_{i_1} + x_{i_2})_{s_t} \, p_{i_1 i_2}$$

where, $p_{i_1 i_2}$ denotes the probability of getting the sample $s_t = (x_{i_1}, x_{i_2})$ in which i_1-th unit occurs at the first draw and i_2-th unit occurs at the second draw. It may be noted that summation is over all possible samples, twelve in number. Collecting the coefficients of x_i for i = 1, 2, 3 and 4

$$\begin{aligned} E(\hat{T}_3) = {} & x_1 [W_{12}p_{12} + W_{21}p_{21} + W_{13}p_{13} + W_{31}p_{31} + W_{14}p_{14} + W_{41}p_{41}] \\ & + x_2 [W_{12}p_{12} + W_{21}p_{21} + W_{23}p_{23} + W_{32}p_{32} + W_{24}p_{24} + W_{42}p_{42}] \\ & + x_3 [W_{13}p_{13} + W_{31}p_{31} + W_{23}p_{23} + W_{32}p_{32} + W_{34}p_{34} + W_{43}p_{43}] \end{aligned}$$

$$+ x_4 \left[W_{14}p_{14} + W_{41}p_{41} + W_{24}\,p_{24} + W_{42}\,p_{42} + W_{34}\,p_{34} + W_{43}\,p_{43}\right]$$

For $\hat{T}_3$ to be an unbiased estimator of the population total we must have

$$E\left(\hat{T}_3\right) = (x_1 + x_2 + x_3 + x_4) \tag{4.3.2}$$

This is to be be true for all $X = (x_1, x_2, x_3, x_4)' \in R^{(4)}$.

Therefore, let us take $X = (1, 0, 0, 0)'$, then (4.3.2) gives the condition

$$W_{12}p_{12} + W_{21}p_{21} + W_{13}\,p_{13} + W_{31}p_{31} + W_{14}p_{14} + W_{41}p_{41} = 1 \tag{4.3.3}$$

Again if we take $X = (0, 1, 0, 0)' \in R^{(4)}$ then (4.3.2) gives another condition

$$W_{12}p_{12} + W_{21}p_{21} + W_{23}p_{23} + W_{32}p_{32} + W_{24}p_{24} + W_{42}p_{42} = 1 \tag{4.3.4}$$

Similarly, by taking $X = (0,0,1,0)'$ and $X = (0,0,0,1)'$ we get the following two more conditions from (4.3.2) viz.

$$W_{31}p_{31} + W_{13}p_{13} + W_{23}p_{23} + W_{32}p_{32} + W_{34}p_{34} + W_{43}p_{43} = 1 \tag{4.3.5}$$

and,

$$W_{14}p_{14} + W_{41}p_{41} + W_{42}p_{42} + W_{24}p_{24} + W_{34}p_{34} + W_{43}p_{43} = 1 \tag{4.3.6}$$

Conversely, if these four equations are true then Eq.(4.3.2) holds. Thus these four conditions are necessary and sufficient conditions to make our estimator unbiased independent of population values. Thus, the four conditions characterise the condition of unbiasedness of $\hat{T}_3$.

In order to prove the consistency of these four equations it is sufficient to show that there exists at least one solution. Let us take

$$W_{i_1 i_2} = \frac{1}{6p_{i_1 i_2}}, \qquad \text{for all } i_1 \text{ and } i_2$$

This set of W's satisfies all the four equations; thereby proving the consistency of the equations.

To obtain MVLUE, we should minimise the variance of $\hat{T}_3$ with respect to the four conditions of unbiasedness using Lagrange's method of multipliers. Let

$$\phi = E\left(\hat{T}_3\right)^2 - T^2 - 2\sum_{i=1}^{4} \lambda_i\,[\,i\,] \tag{4.3.7}$$

where, [i], for i = 1, 2, 3 and 4, represents the Eqs. (4.3.3), (4.3.4), (4.3.5), (4.3.6). We have

$$\begin{aligned} E(\hat{T}_3)^2 = {} & W_{12}^2 (x_1+x_2)^2 p_{12} + W_{13}^2 (x_1+x_3)^2 p_{13} \\ & + W_{14}^2 (x_1+x_4)^2 p_{14} + W_{21}^2 (x_2+x_1)^2 p_{21} \\ & + W_{23}^2 (x_2+x_3)^2 p_{23} + W_{24}^2 (x_2+x_4)^2 p_{24} \\ & + W_{31}^2 (x_3+x_1)^2 p_{31} + W_{32}^2 (x_3+x_2)^2 p_{32} \\ & + W_{34}^2 (x_3+x_4)^2 p_{34} + W_{41}^2 (x_4+x_1)^2 p_{41} \\ & + W_{42}^2 (x_4+x_2)^2 p_{42} + W_{43}^2 (x_4+x_3)^2 p_{43} \end{aligned}$$

Also,

$$\begin{aligned} \sum_{i=1}^{4} \lambda_i [i] = {} & \lambda_1 \{(W_{12}p_{12} + W_{21}p_{21} + W_{13}p_{13} + W_{31}p_{31} + W_{14}p_{14} + W_{41}p_{41}) - 1\} \\ & + \lambda_2 \{(W_{12}p_{12} + W_{21}p_{21} + W_{23}p_{23} + W_{32}p_{32} + W_{24}p_{24} + W_{42}p_{42}) - 1\} \\ & + \lambda_3 \{(W_{13}p_{13} + W_{31}p_{31} + W_{23}p_{23} + W_{32}p_{32} + W_{34}p_{34} + W_{43}p_{43}) - 1\} \\ & + \lambda_4 \{(W_{14}p_{14} + W_{41}p_{41} + W_{24}p_{24} + W_{42}p_{42} + W_{34}p_{34} + W_{43}p_{43}) - 1\} \end{aligned}$$

Equating $\partial\phi / \partial W_{12} = 0$ we get

$$2W_{12}(x_1+x_2)^2 p_{12} - 2\lambda_1 p_{12} - 2\lambda_2 p_{12} = 0$$

Therefore,

$$W_{12} = \frac{\lambda_1 + \lambda_2}{(x_1+x_2)^2} \tag{4.3.8}$$

Equating $\partial\phi / \partial W_{21} = 0$, we have $W_{21} = \dfrac{\lambda_1 + \lambda_2}{(x_1+x_2)^2}$

Similarly,

$$\frac{\partial\phi}{\partial W_{13}} = 0 \quad \Rightarrow \quad W_{13} = \frac{\lambda_1 + \lambda_3}{(x_1+x_3)^2}$$

$$\frac{\partial\phi}{\partial W_{14}} = 0 \quad \Rightarrow \quad W_{14} = \frac{\lambda_1 + \lambda_4}{(x_1+x_4)^2}$$

$$\frac{\partial\phi}{\partial W_{23}} = 0 \quad \Rightarrow \quad W_{23} = \frac{\lambda_2 + \lambda_3}{(x_2+x_3)^2}$$

$$\frac{\partial\phi}{\partial W_{24}} = 0 \quad \Rightarrow \quad W_{24} = \frac{\lambda_2 + \lambda_4}{(x_2 + x_4)^2}$$

$$\frac{\partial\phi}{\partial W_{34}} = 0 \quad \Rightarrow \quad W_{34} = \frac{\lambda_3 + \lambda_4}{(x_3 + x_4)^2}$$

Differentiating ϕ with respect to $W_{31}, W_{41}, W_{32}, W_{42}$ and W_{43} and equating them to zero we note as in Eq. (4.3.8)

$$W_{31} = W_{13}\ ; \quad W_{32} = W_{23}\ ; \quad W_{41} = W_{14}\ ;$$

$$W_{42} = W_{24}; \quad W_{43} = W_{34}.$$

Further, we note that

$$\begin{aligned} W_{12}(x_1 + x_2)^2 - W_{13}(x_1 + x_3)^2 &= \lambda_1 + \lambda_2 - (\lambda_1 + \lambda_3) \\ &= \lambda_2 - \lambda_3 \end{aligned}$$

and also

$$\begin{aligned} W_{34}(x_3 + x_4)^2 - W_{24}(x_2 + x_4)^2 &= \lambda_3 + \lambda_4 - \lambda_2 - \lambda_4 \\ &= \lambda_3 - \lambda_2 \end{aligned}$$

Therefore,

$$W_{12}(x_1 + x_2)^2 - W_{13}(x_1 + x_3)^2 + W_{34}(x_3 + x_4)^2 - W_{24}(x_2 + x_4)^2 = 0 \quad (4.3.9)$$

Let the coefficients W's be independent of x's. Then the LHS of Eq. (4.3.9) is a quadratic form of type $X'BX$. Therefore, by Lemma 4.2.1.

$$W_{12} = W_{13} = W_{34} = W_{24} = 0 \quad (4.3.10)$$

Again, if we consider

$$W_{14}(x_1 + x_4)^2 + W_{23}(x_2 + x_3)^2 - W_{24}(x_2 + x_4)^2 - W_{13}(x_1 + x_3)^2 = 0 \quad (4.3.11)$$

The LHS of Eq. (4.3.11) is also a quadratic form of the type $X'BX$. Therefore, again by Lemma (4.2.1)

$$W_{14} = W_{23} = W_{24} = W_{13} = 0 \quad (4.3.12)$$

Using the set of W's in Eqs. (4.3.10) and (4.3.12) and noting that $W_{ij} = W_{ji}$ for all i, j. We have

$$\begin{aligned} W_{12} &= W_{13} = W_{14} = W_{21} = W_{23} = W_{24} = W_{31} = W_{32} \\ &= W_{34} = W_{41} = W_{42} = W_{43} = 0 \end{aligned} \quad (4.3.13)$$

Thus we prove that all $W_{i_1 i_2}$ are zero. However, these values of W's do not satisfy the set of Eqs. (4.3.3), (4.3.4), (4.3.5) and (4.3.6), characterising the condition of unbiasedness. Thus, the assumption that W's do not depend on X leads to a contradiction. This proves that the MVLUE, independent of population values, in T_3 – class of linear estimators does not exist for this special case.

Remark : If in the proof of the above theorem, we relax the condition that W's should be independent of x's and this may depend on x's, then for simple random sampling without replacement, the coefficients W's are given as follows :

$$W_{12} = W_{21} = W_{34} = W_{43} = \frac{k}{(x_1 + x_2)^2 + (x_3 + x_4)^2};$$

$$W_{13} = W_{31} = W_{24} = W_{42} = \frac{k}{(x_1 + x_3)^2 + (x_2 + x_4)^2}$$

and,

$$W_{14} = W_{41} = W_{23} = W_{32} = \frac{k}{(x_1 + x_4)^2 + (x_2 + x_3)^2}$$

where,

$$k = (\lambda_1 + \lambda_2 + \lambda_3 + \lambda_4) = 6\left[\frac{1}{(x_1 + x_2)^2 + (x_3 + x_4)^2} + \frac{1}{(x_1 + x_3)^2 + (x_2 + x_4)^2} + \frac{1}{(x_1 + x_4)^2 + (x_2 + x_3)^2}\right]^{-1} \quad (4.3.14)$$

We note that the MVLUE estimator, for this sampling scheme, is an unordered estimator involving all the values of the population. Thus, this estimator is no serviceable, unlike the regression and double sampling estimators in T_1 – class, discussed in Chapter VI of Lecture Notes (Tikkiwal, 1985), where the regression coefficient occurring in the estimators also depends on population values to begin with, but it is estimable from the sample. Thus, the examination of non-existence of MVLUE estimator in a class independent of population values, may mislead us in certain cases. For more details, a reader may refer to a paper by the author and Tikkiwal, G.C. (1979).

General Proof :

Let a sample of size n be drawn from a population of size N with sampling without replacement. Then the number of possible samples when order is taken into consideration is $n!\binom{N}{n}$. In the s_t-th sample for t = 1, 2, . . ., $n!\binom{N}{n}$, let the unit selected at the first draw be denoted by i_1, at the second draw by i_2 and so on. Thus $i_1, i_2, \ldots, i_n$ are some n units of the population in the sample s_t. Further, let $p_{i_1 i_2 \ldots i_n}$ be the probability p_{s_t} of selecting the s_t-th sample and $\beta^{s_t} = W_{i_1 i_2 \ldots i_n}$ be the weight associated to the total of this sample. The weight depends, in general, on the order of the units and thus for given n units there are n! weights. By definition, the general estimator in the T_3-class is

$$\hat{T}_3 = \beta^{s_t} (\Sigma X_r)_{s_t} \tag{4.3.15}$$

where, X_r denotes the variate value of the unit drawn at r-th draw. Therefore,

$$E(\hat{T}_3) = \sum_{t=1}^{n!\binom{N}{n}} \beta^{s_t} \left[\sum_{r=1}^{n} X_r\right] p_{s_t}$$

$$= \sum_{t=1}^{n!\binom{N}{n}} W_{i_1 i_2 \ldots i_n} \left[\sum_{r=1}^{n} x_{i_r}\right] p_{i_1 i_2 \ldots i_n} \tag{4.3.16}$$

By collecting the coefficients of x_i's

$$\begin{aligned} E(\hat{T}_3) = x_1 & \left[\Sigma' W_{1 i_2 \ldots i_n} \cdot p_{1 i_2 \ldots i_n} + \Sigma' W_{i_1 1 i_3 \ldots i_n} \cdot p_{i_1 1 i_3 \ldots i_n} + \cdots \right. \\ & \left. + \Sigma' W_{i_1 i_2 \ldots i_{n-1} 1} \cdot p_{i_1 i_2 \ldots i_{n-1} 1}\right] \\ & + x_2 \left[\Sigma' W_{2 i_2 \ldots i_n} \cdot p_{2 i_2 \ldots i_n} + \Sigma' W_{i_1 2 i_3 \ldots i_n} \cdot p_{i_1 2 i_3 \ldots i_n} + \cdots \right. \\ & \left. + \Sigma' W_{i_1 i_2 \ldots i_{n-1} 2} \cdot p_{i_1 i_2 \ldots i_{n-1} 2}\right] + \ldots \\ & + x_N \left[\Sigma' W_{N i_2 \ldots i_n} \cdot p_{N i_2 \ldots i_n} + \Sigma' W_{i_1 N i_3 \ldots i_n} \cdot p_{i_1 N i_3 \ldots i_n} + \ldots \right. \\ & \left. + \Sigma' W_{i_1 i_2 \ldots i_{n-1} N} \cdot p_{i_1 i_2 \ldots i_{n-1} N}\right] \end{aligned} \tag{4.3.17}$$

where, Σ' stands for summation over all possible samples $s_t = (i_1, i_2, \ldots, i_{r-1}, j, i_{r+1}, \ldots, i_n)$ in which a particular j-th unit occurs at a particular r-th draw, for j = 1, 2, . . ., N and r = 1, 2, . . ., n. Such possible samples are $(n-1)!\binom{N-1}{n-1}$ in number. For $\hat{T}_3$ to be unbiased estimator of the population total T, from Eq. (4.3.17) we must have for $j = 1, 2, \ldots, N$;

$$\sum_{r=1}^{n} \Sigma' W_{i_1 i_2 \ldots i_{r-1} j i_{r+1} \ldots i_n} \cdot p_{i_1 i_2 \ldots i_{r-1} j i_{r+1} \ldots i_n} = 1 \tag{4.3.18}$$

Thus, as in special case, we have N equations which characterise the unbiasedness of $\hat{T}_3$ for all $X = (x_1, x_2, \ldots, x_N)$. It will be sufficient to obtain a solution to show consistency of the set of N equations given in Eq. (4.3.18).

Let $W_{i_1 i_2 \ldots i_{r-1} j i_{r+1} \ldots i_n}$

$$= \frac{1}{n!\binom{N-1}{n-1}} \cdot \frac{1}{p_{i_1 i_2 \ldots i_{r-1} j i_{r+1} \ldots i_n}} \quad \text{for } j = 1, 2, \ldots, N \tag{4.3.19}$$

This set of W's in Eq. (4.3.19) satisfies all the N equations in Eq. (4.3.18) and hence this set of equations is consistent. Thus, T_3 – class is non-empty for sampling without replacement.

Now we proceed to examine the existence of MVLUE. For this we construct the following function

$$\phi = \text{Var}\left(\hat{T}_3\right) - 2 \sum_{j=1}^{N} \lambda_j \, [j] \tag{4.3.20}$$

where [j] stands for the Eq. (4.3.18) and λ_j is the Lagrange's multiplier .

$$\phi = E\left(\hat{T}_3\right)^2 - T^2 - 2 \sum_{j=1}^{N} \lambda_j \, [j]$$

$$= \sum_{s_t=1}^{n!\binom{N}{n}} W_{i_1 i_2 \ldots i_n}^2 \left[\sum_{r=1}^{n} x_{i_r}\right]^2 \cdot p_{i_1 i_2 \ldots i_n} - T^2$$

$$-2 \sum_{j=1}^{N} \lambda_j \sum_{r=1}^{n} W_{i_1 i_2 \ldots i_{r-1} j i_{r+1} \ldots i_n} \cdot p_{i_1 i_2 \ldots i_{r-1} j i_{r+1} \ldots i_n} \tag{4.3.21}$$

Equating $\dfrac{\partial \phi}{\partial W_{i_1 i_2 \ldots i_{r-1} j i_{r+1} \ldots i_n}} = 0$ and simplifying,

We get

$$W_{i_1 i_2 \ldots i_{r-1} j i_{r+1} \ldots i_n} = \frac{\lambda_{i_1} + \lambda_{i_2} + \lambda_{i_{r-1}} + \lambda_j + \lambda_{i_{r+1}} + \ldots + \lambda_{i_n}}{(x_{i_1} + x_{i_2} + \ldots + x_{i_{r-1}} + x_j + x_{i_{r+1}} + \ldots + x_{i_n})^2} \qquad (4.3.22)$$

Thus Eq. (4.3.22) gives the values of different W's. It is evident that these W's do not depend upon the order of the units in which they appear in the sample and thus the estimator $\hat{T}_3$ so obtained would be unordered. Let us consider the following four sets of W's from Eq. (4.3.22).

(1) $$W_{i_1 i_2 \ldots i_n} = \frac{\lambda_{i_1} + \lambda_{i_2} + \ldots + \lambda_{i_n}}{(x_{i_1} + x_{i_2} + \ldots + x_{i_n})^2}$$

(2) $$W_{j_1 i_2 \ldots i_n} = \frac{\lambda_{j_1} + \lambda_{i_2} + \ldots + \lambda_{i_n}}{(x_{j_1} + x_{i_2} + \ldots + x_{i_n})^2}$$

(3) $$W_{j_1 j_2 \ldots j_n} = \frac{\lambda_{j_1} + \lambda_{j_2} + \ldots + \lambda_{j_n}}{(x_{j_1} + x_{j_2} + \ldots + x_{j_n})^2}$$

(4) $$W_{i_1 j_2 \ldots j_n} = \frac{\lambda_{i_1} + \lambda_{j_2} + \lambda_{j_3} + \ldots + \lambda_{j_n}}{(x_{i_1} + x_{j_2} + x_{j_3} + \ldots + x_{j_n})^2}$$

where, none of the units being common to the two samples $(i_1 i_2 \ldots i_n)$ and $(j_i\, j_2 \ldots j_n)$.

From the above four equations, we obtain

$$W_{i_1 i_2 \ldots i_n}(x_{i_1} + x_{i_2} + \ldots + x_{i_n})^2 - W_{j_1 j_2 \ldots j_n}(x_{j_1} + x_{j_2} + \ldots + x_{j_n})^2$$
$$= \lambda_{i_1} - \lambda_{j_1}$$
$$= W_{i_1 j_2 \ldots j_n}(x_{i_1} + x_{j_2} + \ldots + x_{j_n})^2 - W_{j_1 j_2 \ldots j_n}(x_{j_1} + x_{j_2} + \ldots + x_{j_n})^2$$

or,

$$W_{i_1 i_2 \ldots i_n}(x_{i_1} + x_{i_2} + \ldots + x_{i_n})^2 - W_{j_1 i_2 \ldots i_n}(x_{j_1} + x_{i_2} + \ldots + x_{i_n})^2$$
$$+ W_{j_1 j_2 \ldots j_n}(x_{j_1} + x_{j_2} + \ldots + x_{j_n})^2 - W_{i_1 j_2 \ldots j_n}(x_{i_1} + x_{j_2} + x_{j_3} + \ldots + x_{j_n})^2$$
$$= 0 \qquad (4.3.23)$$

Assuming W's to be independent of x's, the Eq. (4.3.23) can be expressed as $X' BX = 0$ where $X = (x_{i_1}, x_{i_2}, \ldots, x_{i_n}, x_{j_1}, x_{j_2}, \ldots, x_{j_n})$ and B is the matrix of the quadratic form involving the W's.

Now using Lemma (4.2.1) we get

$$W_{i_1 i_2 \ldots i_n} = W_{j_1 j_2 \ldots j_n} = W_{j_1 i_2 \ldots i_n} = W_{i_1 j_2 \ldots j_n} = 0$$

Considering the relevant equations from Eq. (4.3.22) and making use of symmetry we get all the W's equal to zero. But this contradicts the conditions of unbiasedness in Eq. (4.3.18). Thus the MVLUE independent of population value does not exist in T_3-class of estimation, when sampling without replacement is used.

4.4. AN ESTIMATOR IN T_3 – CLASS FOR MIDZUNO SCHEME OF SAMPLING WITHOUT REPLACEMENT

We know that in Midzuno scheme of sampling without replacement (See Section 2.2 (b) of Chapter 2), first draw is made with varying probabilities and the subsequent (n–1) draws for n > 1 are performed with equal probabilities and without replacement. In this sampling scheme we associate the weight $\beta^{s_t} = W_{i_1}$ dependent on the unit occurring in the s_t-th sample at the first draw i.e. when $X_1 = x_{i_1}$. Then the Midzuno's estimator in T_3 – class for sampling without replacement is

$$\hat{T}_3 = W_{i_1} \left(\sum_{r=1}^{n} x_{i_r} \right) \tag{4.4.1}$$

It is shown in the literature (Prabhu Ajgaonkar, 1962) that there is only one unbiased estimator of this type and so it is MVLUE. The value of W_{i_1} which makes the estimator unbiased is given by

$$W_{i_1} = \frac{1}{np_{i_1}} \tag{4.4.2}$$

where, p_{i_1} is the probability of selecting a particular unit i_1 at the first draw. The variance of the estimator in Eq. (4.4.1) is given by

$$\text{Var}\left(\hat{T}_3\right) = E\left(\hat{T}_3\right)^2 - T^2 \tag{4.4.3}$$

where,

$$E\left(\hat{T}_3\right)^2 = \sum_{i=1}^{N} \frac{n-1}{n^2(N-1)}\left\{S + \frac{N-n}{(n-1)\,p_i}\right\} x_i^2$$

$$+ \sum_{i \neq j}^{N} \frac{n-1}{n^2(N-1)}\left\{\frac{n-2}{N-2} S + \frac{N-n}{N-2} \cdot \left(\frac{1}{p_i} + \frac{1}{p_j}\right)\right\} x_i x_j \qquad (4.4.4)$$

with $\quad S = \sum_{i=1}^{N} (1/p_i)$.

If $s_t = (i_1, i_2, \ldots, i_n)$, then

$$p_{s_t} = p_{i_1} \cdot \frac{1}{N-1} \cdot \frac{1}{N-2} \cdots \frac{1}{N-n+1}$$

$$= p_{i_1} \cdot \frac{(N-n)!}{(N-1)!}$$

We note that $\hat{T}_3$ is an ordered estimator. The unordered estimator, for a given set of $(i_1, i_2, \ldots, i_n)$ units occurring in n! samples, is

$$\hat{T}_3' = \frac{\Sigma^{(1)}\Sigma^{(2)} P\left\{s_t\,(1, i_j)\right\}\left\{\left(\sum_{j=1}^{n} x_{i_j}\right)\Big/ np_{i_j}\right\}}{\left[\Sigma^{(1)}\Sigma^{(2)} P\left\{s_t\,(1, i_j)\right\}\right]} \qquad (4.4.5)$$

where, $\Sigma^{(2)}$ is the sum over all those $(n-1)!$ samples $s_t(1, i_j)$ in which i_j-th unit occurs at the first draw and remaining (n–1) units in some given permutation. $\Sigma^{(1)}$ is the sum over j = 1,2,. . ., n. The numerator of the RHS of Eq. (4.4.5) is

$$\frac{1}{n}\Sigma^{(1)}\Sigma^{(2)} p_{i_j} \frac{(N-n)!}{(N-1)!} \cdot \frac{1}{p_{i_j}} \cdot \left[\sum_{j=1}^{n} x_{i_j}\right]$$

$$= \frac{1}{n}\Sigma^{(1)} \frac{(N-n)!}{(N-1)!}(n-1)!\left(\sum_{j=1}^{n} x_{i_j}\right)$$

$$= n\,\frac{(n-1)!}{n} \cdot \frac{(N-n)!}{(N-1)!}\left(\sum_{j=1}^{n} x_{i_j}\right)$$

$$= \frac{(n-1)!}{(N-1)!}(N-n)! \sum_{j=1}^{n} x_{i_j}$$

Further the denominator of R.H.S. is

$$\frac{(n-1)!\,(N-n)!}{(N-1)!}\sum_{j=1}^{n} p_{i_j} = \left\{1/\binom{N-1}{n-1}\right\}\sum_{j=1}^{n} p_{i_j} = p'_{s_t}$$

where p'_{s_t} is the probability of getting $(i_1, i_2, \ldots, i_n)$ units in the samples s_t in different orders. Therefore,

$$\hat{T}'_3 = \frac{\sum_{j'=1}^{n} x_{i_{j'}}}{\sum_{j'=1}^{n} p_{i_{j'}}} \tag{4.4.6}$$

The above estimator reduces to the well known ratio estimator, if

$$p_{i_j} = \frac{y_{i_j}}{\sum_{j=1}^{N} y_{i_j}}$$

where y_{i_j} is the size of the i_j-th unit for j = 1, 2, . . ., n. The variance of this unordered estimator $\hat{T}'_3$ is given by

$$\mathrm{Var}\left(\hat{T}'_3\right) = \frac{1}{\binom{N-1}{n-1}^2}\left[\sum_{t=1}^{\binom{N}{n}}\left\{\sum_{j'=1}^{n} x_{i_{j'}}\right\}^2 / p'_{s_t}\right] - T^2$$

$$= \frac{1}{\binom{N-1}{n-1}}\left[\sum_{t=1}^{\binom{N}{n}}\left\{\frac{\left(\sum_{j'} x_{i_{j'}}\right)^2}{\sum_{j'} p_{i_{j'}}}\right\}\right] - T^2 \tag{4.4.7}$$

[Tikkiwal, 1966, Eq. (1.14), p. 67].

That even the unordered estimator $\hat{T}'_3$ based on the varying probability sampling scheme due to Midzuno can be worse, in the sense of efficiency, than SRSWOR estimator is shown by Tikkiwal (1966). Thus, there is need for caution in using varying probability estimators indiscriminantly. This point is also supported from the discussions in Chapter 2 on T_1-class and will be further supported from the discussions in Chapter 6.

4.5 NON-EXISTENCE OF MVLUE IN THE CLASS FOR SAMPLING WITH REPLACEMENT

We have seen that the MVLUE does not exist in T_3 – class of estimators, but it does exist in a sub-class for Midzuno sampling scheme without replacement. In this and the following sections we take the case when sampling is carried out with replacement.

Let a sample of size n be drawn with varying probabilities and with replacement from a population of size N. Then, the number of possible samples are N^n. Further a particular sample s_t may or may not have all distinct n units. Let p_{s_t} be given by $p_{i_1 i_2 \ldots i_n}$ and $W_{i_1 i_2 \ldots i_n}$ be the weight associated with this s_t-th sample as in Section 4.3. Then, by definition, T_3 – class estimator under SWR is

$$\hat{T}_3 = W_{i_1 i_2 \ldots i_n} \left[\sum_{r=1}^{n} x_{i_r} \right] \qquad (4.5.1)$$

Then,

$$E\left(\hat{T}_3\right) = \sum_{t=1}^{N^n} W_{i_1 i_2 \ldots i_n} \left[\sum_{r=1}^{n} x_{i_r} \right]_{s_t} p_{i_1 i_2 \ldots i_n} \qquad (4.5.2)$$

Collecting the coefficients of individual x_i 's in Eq. (4.5.2)

$$\begin{aligned} E\left(\hat{T}_3\right) = x_1 \Bigg[& n \,.\, W_{111\ldots 1}\, p_{111\ldots 1} + (n-1) \sum_{i_1 (\neq 1)}^{N-1} W_{i_1 11\ldots 1}\, p_{i_1 11\ldots 1} \\ & + \ldots + (n-1) \sum_{i_n (\neq 1)}^{N-1} W_{111\ldots i_n}\, p_{111\ldots i_n} \\ & + (n-2) \sum_{i_1 i_2 (\neq 1)}^{(N-1)^2} W_{i_1 i_2 1\ldots 1}\, p_{i_1 i_2 1 \ldots 1} + \ldots \\ & + (n-2) \sum_{i_{n-1} i_n (\neq 1)}^{(N-1)^2} W_{111\ldots i_{n-1} i_n}\, p_{111\ldots i_{n-1} i_n} \\ & + \ldots + \sum_{i_2 i_3 \ldots i_n (\neq 1)}^{(N-1)^{n-1}} W_{1 i_2 \ldots i_n} \cdot p_{1 i_2 \ldots i_n} \Bigg] \\ & + x_2 [\quad] + x_3 [\quad] + x_N [\quad] \end{aligned} \qquad (4.5.3)$$

The coefficients for $x_2, x_3, \ldots, x_N$ can be obtained on the lines of those for x_1. The number of times in which a particular unit x_j occurs in N^n samples is equal to

$$n+\binom{n}{1}(n-1)(N-1)+\binom{n}{2}(n-2)(N-1)^2+\ldots+\binom{n}{n-1}\{n-(n-1)\}(N-1)^{n-1}$$

$$= n\,[1+(N-1)]^{n-1}$$

$$= n\,.\,N^{n-1} \tag{4.5.4}$$

In order that the estimator $\hat{T}_3$ in Eq. (4.5.1) be unbiased estimator of the population total T for all X,

$$n\,W_{j\ldots j}\,p_{j\ldots j}+(n-1)\sum_{i_1(\neq j)}^{(N-1)} W_{i_1 jj\ldots j}\,p_{i_1 jj\ldots j}$$

$$\ldots+\sum_{i_2 i_3\ldots i_n(\neq j)}^{(N-1)^{n-1}} W_{j i_2 i_3\ldots i_n}\,p_{j i_2 i_3\ldots i_n}=1 \tag{4.5.5}$$

for j= 1, 2, . . . , N. Clearly Eq. (4.5.5) gives a set of N equations in N^n W's characterising the condition of unbiasedness of $\hat{T}_3$ for all $X=(x_1, x_2, \ldots, x_N)$. The above set of equations in Eq. (4.5.5) are consistent if there exists at least one set of W's satisfying the above equations. It can be easily seen that the set of W's with $W_{i_1 i_2\ldots i_n}=nN^{n-1}/p_{i_1 i_2\ldots i_n}$ satisfies all the N equations in (4.5.5) and so T_3-class is non-empty for sampling with replacement.

The variance of the estimator $\hat{T}_3$ is given by

$$\text{Var}\left(\hat{T}_3\right)=E\left(\hat{T}_3\right)^2-T^2 \tag{4.5.6}$$

where,

$$E\left(\hat{T}_3^2\right)=\sum_{s_t=1}^{\binom{N^n}{}} W_{i_1 i_2\ldots i_n}^2\left[\sum_{r=1}^{n} x_{i_r}\right]_{s_t}^2 \cdot p_{i_1 i_2\ldots i_n}$$

The function to determine the optimum set of W's is

$$\phi=\text{Var}\left(\hat{T}_3\right)-2\sum_{j=1}^{N}\lambda_j\;[\,j\,] \tag{4.5.7}$$

where, λ_j is Lagrange's multiplier corresponding to the j-th Eq. (4.5.5) and [j] represents the j-th equation, for j=1,2, . . ., N. The minimization of ϕ leads to

$$W_{i_1 i_2 \ldots i_n} = \frac{\lambda_{i_1} + \lambda_{i_2} + \ldots + \lambda_{i_n}}{(x_{i_1} + x_{i_2} + \ldots + x_{i_n})^2} \tag{4.5.8}$$

It is clear from Eq. (4.5.8) that the weights are indepedent of the order in which the units occur in a sample and so the estimator based on these weights is unordered. We consider the following weights flowing from the above equation

$$W_{jj\ldots j} = \frac{n\lambda_j}{(nx_j)^2}$$

$$W_{kk\ldots k} = \frac{n\lambda_k}{(nx_k)^2}$$

$$W_{i_1 i_2 \ldots i_{r-1}\, j\, i_{r+1} \ldots i_n} = \frac{\lambda_{i_1} + \lambda_{i_2} + \ldots + \lambda_{i_{r-1}} + \lambda_j + \lambda_{i_{r+1}} + \ldots + \lambda_{i_n}}{(x_{i_1} + x_{i_2} + \ldots + x_{i_{r-1}} + x_j + x_{i_{r+1}} + \ldots + x_{i_n})^2}$$

and,

$$W_{i_1 \ldots i_{r-1}\, k\, i_{r+1} \ldots i_n} = \frac{\lambda_{i_1} + \lambda_{i_2} + \ldots + \lambda_{i_{r-1}} + \lambda_k + \lambda_{i_{r+1}} + \ldots + \lambda_{i_n}}{(x_{i_1} + x_{i_2} + \ldots + x_{i_{r-1}} + x_k + x_{i_{r+1}} + \ldots + x_{i_n})^2}$$

From this set of four W's we obtain

$$n W_{jj\ldots j}\, x_j^2 + W_{i_1 \ldots i_{r-1}\, k\, i_{r+1} \ldots i_n} (x_{i_1} + x_{i_2} + \ldots + x_{i_{r-1}} + x_k + x_{i_{r+1}} + x_{i_n})^2$$

$$- n W_{kk\ldots k}\, x_k^2 - W_{i_1 i_2 \ldots i_{r-1}\, j\, i_{r+1} \ldots i_n} (x_{i_1} + x_{i_2} + \ldots + x_{i_{r-1}} + x_j + x_{i_{r+1}} + x_{i_n})^2$$

$$= 0 \tag{4.5.9}$$

Assuming weights to be independent of x's, we can write Eq. (4.5.9) in the form

$X'BX = 0$ where, $X' = (x_{i_1}, x_{i_2}, \ldots, x_{i_{r-1}}, x_k, x_{i_{r+1}}, \ldots, x_{i_n}, x_j)$

Therefore, by Lemma (4.2.1) we find that

$$W_{jj\ldots j} = W_{kk\ldots k} = W_{i_1 i_2 \ldots i_{r-1}\, k\, i_{r+1} \ldots i_n}$$

$$= W_{i_1 i_2 \ldots i_{r-1}\, j\, i_{r+1} \ldots i_n} = 0 \tag{4.5.10}$$

Similarly we can also prove that other W's are also zero; contradicting the condition of unbiasedness. Thus W's depend upon X.

Therefore, there does not exist MVLUE independent of population values in T_3-class for SWR.

4.6 AN ESTIMATOR IN THE CLASS FOR MIDZUNO SCHEME OF SAMPLING WITH REPLACEMENT

In this scheme of sampling, a unit at the first draw is selected with varying probabilities and the subsequent (n–1) units for n > 1 are selected with equal probabilities and with replacement.

Let p_i be the probability of selecting the i-th unit at the first draw for i=1,2,. . . , N and further let p_{s_t} be the probability of the s_t-th sample in which the unit drawn at the first draw is i_1. Then

$$p_{s_t} = p_{i_1} \cdot \frac{1}{N^{n-1}}$$

Let the estimator of the population total under Midzuno scheme of sampling with replacement in T_3-class be

$$\hat{T}_3 = W_{i_1}\left[\sum_{r=1}^{n} x_{i_r}\right] \tag{4.6.1}$$

where, the weight W_{i_1} depends upon the outcome at the first draw. Thus the weights associated with the samples, having the same element at the first draw, are the same. In general there will be N distinct values of W_{i_1} for $i_1 = 1, 2, \ldots, N$. It is shown [Prabhu Ajgaonkar, (1962)] that there is a unique unbiased estimator for $W_{i_1} = 1/np_{i_1}$, and so it is MVLUE in this sub-class of T_3. Thus, the MVLUE in this sub-class is

$$\hat{T}_3 = \frac{1}{np_{i_1}}\left[\sum_{r=1}^{n} x_{i_r}\right] \tag{4.6.2}$$

The variance of this estimator is

$$V\left(\hat{T}_3\right) = \sum_{i=1}^{N} \frac{x_i^2}{Nn^2}\left[\frac{S(n-1)(N+n-1)}{N} + \frac{N+2(n-1)}{p_i}\right]$$

$$+ \sum_{i=1}^{N} \sum_{j(\neq i)}^{N} \frac{(n-1)\, x_i\, x_j}{Nn^2}\left[\frac{(n-2)S}{N} + \left(\frac{1}{p_i} + \frac{1}{p_j}\right)\right] \tag{4.6.3}$$

where, $S = \sum_{i=1}^{N} (1/p_i)$

One can now obtain the unordered and combined unordered estimators of $\hat{T}_3$ given in Eq. (4.6.2), using results of Sections 1.4 and 1.6 of Chapter 1.

4.7 A UNIFIED PROOF OF NON-EXISTENCE OF MVLUE IN THE CLASS FOR SAMPLING WITH OR WITHOUT REPLACEMENT

The following unified proof is due to Bhargava (1975). The estimator $\hat{T}_3$ in Eq. (4.1.1) can be written as

$$\hat{T}_3 = \beta^{s_t} \left(\sum_{j=1}^{k} R_{i_j}^{(p)} x_{i_j} \right)_{s_t} \qquad (4.7.1)$$

where, $R_{i_j}^{(p)}$ denotes the number of times i_j-th unit occurs in the sample s_t for $j=1,2,\ldots, k$ for

$$s_t \in S_p \left\{ \left(i_1, R_{i_1}^{(p)} \right), \ldots, \left(i_k, R_{i_k}^{(p)} \right) \right\}$$

in the notations of Section 1.1 of Chapter 1. Then,

$$E\left(\hat{T}_3\right) = \sum_{s_t \in S'} \left[\beta^{s_t} \left(\sum_{j=1}^{k} R_{i_j}^{(p)} x_{i_j} \right)_{s_t} \right] p(s_t)$$

where, S' is the sample space

$$= \sum_{k=1}^{n} \sum_{l=1}^{\binom{N}{k}} \sum_{p=1}^{M_{kl}} \sum_{t=1}^{M_{klp}} \beta^{s_t} \left(\sum_{j=1}^{k} R_{i_j}^{(p)} x_{i_j} \right)_{s_t} p(s_t)$$

$$= \sum_{i=1}^{N} x_i \left[\sum_{k=1}^{n} \sum_{l=1}^{\binom{N-1}{k-1}} \sum_{p=1}^{M_{kl}} \sum_{t=1}^{M_{klp}} \beta^{s_t} R_i^{(p)} p(s_t) \right]$$

The necessary and sufficient conditions for $\hat{T}_3$ to be an unbiased estimator of T, for all $X \in R^{(N)}$, is

$$\sum_{k=1}^{n} \sum_{l=1}^{\binom{N-1}{k-1}} \sum_{p=1}^{M_{kl}} \sum_{t=1}^{M_{klp}} \beta^{s_t} R_i^{(p)} p(s_t) = 1 \qquad (4.7.2)$$

with $s_t \supset i$-th unit for $i = 1, 2, \ldots, N$.

Let C(i) denote the total number of times the i-th unit occurs in different samples $s_t \in S'$. Then

$$C(i) = \sum_{k=1}^{n} \sum_{l=1}^{\binom{N-1}{k-1}} \sum_{p=1}^{M_{kl}} \sum_{t=1}^{M_{klp}} \left(R_i^{(p)}\right) \tag{4.7.3}$$

Since, for sampling without replacement, $C(i) = n!\binom{N-1}{n-1}$ and for sampling with replacement $= n\,N^{n-1}$ from Eq. (4.5.4), we can take $C(i) = C$, a constant quantity, for all (i). Let $\beta^{s_t} = 1/\{C\,p\,(s_t)\}$ for all $s_t \in S'$. Such a set of β's satisfies the N equations characterising the condition of unbiasedness. Therefore, these equations are consistent and T_3 – class is thus non-empty. Now,

$$E\left(\hat{T}_3^2\right) = \sum_{k=1}^{n} \sum_{l=1}^{\binom{N}{k}} \sum_{p=1}^{M_{kl}} \sum_{t=1}^{M_{klp}} \left[\beta^{s_t^{\,2}} \left(\sum_{j=1}^{k} R_{i_j}^{(p)} x_{i_j}\right)_{s_t}^{2} p(s_t)\right]$$

For obtaining optimum values of β^{s_t}, let

$$\phi = V\left(\hat{T}_3\right) - 2\sum_{j=1}^{N} \lambda_j\,[j]$$

as in Section 4.5. Equating $\left(\partial\phi/\,\partial\beta^{s_t}\right) = 0$ for a given s_t, we get

$$\beta^{s_t}\left(\sum_{j=1}^{k} R_{i_j}^{(p)} x_{i_j}\right)_{s_t} = \sum_{j=1}^{k} \lambda_{i_j} R_{i_j}^{(p)} \tag{4.7.4}$$

Since we require MVLUE to be independent of population values and so the β's, consider first a population for which all $x_{i_j} \in s_t$ are zero, then

$$\sum_{j=1}^{k} \lambda_{i_j} R_{i_j}^{(p)} = 0$$

Let us now consider a population in which none of $x_{i_j} \in s_t$ are zero. Then β^{s_t} from Eq. (4.7.4) is zero for that s_t. Since this argument applies to all $s_t \in S'$; thus

$$\beta^{s_t} = 0, \qquad \text{for all } s_t \in S' \tag{4.7.5}$$

But, this solution for β's contradicts the condition of unbiasedness. Hence, MVLUE in T_3 – class does not exist for sampling with or without replacement.

EXERCISES

4.1. Under the Midzuno scheme of sampling without replacement, consider the following estimator of population total T in T_3 -class

$$\hat{T}_3 = W_{i_1}\left[\sum_{r=1}^{n} (X_r = x_{i_r})\right] \tag{1}$$

where, W_{i_1}, the weight associated with the sample, depends on the out-come of the first draw x_{i_1}, i.e. when $X_1 = x_{i_1}$. Show that $\hat{T}_3$ is MVLUE for the value

$$W_{i_1} = \frac{1}{n\, p_{i_1}} \tag{2}$$

where p_{i_1} is the probability of selecting a particular unit i_1 at the first draw. Also obtain expression of the variance of $\hat{T}_3$.

4.2. Under the Midzuno scheme of sampling without replacement the estimator given in Eq. (1) of Exercise 4.1 is an ordered estimator. Show that the unordered estimator, for a given set of $(i_1, i_2, \ldots, i_n)$ units occurring in n! samples, is

$$\hat{T}_{uo} = \frac{\sum\limits_{j=1}^{n} x_{i_j}}{\sum\limits_{j=1}^{n} p_{i_j}} \tag{3}$$

Further, show that

$$V\left(\hat{T}_{uo}\right) \le V\left(\hat{T}_3\right).$$

In which calss of linear estimators does $\hat{T}_{uo}$ belong ?

4.3. Show that under the Midzuno scheme of sampling WOR the unordered estimator $\hat{T}_{uo}$ given in Eq. (3) of Exercise 4.2 reduces to the classical ratio estimator

$$\hat{T}_r = \frac{\overline{X}_n}{\overline{Y}_n}\, \overline{y}_N \tag{4}$$

for

$$p_{i_j} = \frac{y_{i_j}}{\sum_{j=1}^{N} y_{i_j}}$$

where, y_{i_j} is the size of the i_j-th unit for j = 1, 2, . . . , n; and $\overline{X}_n, \overline{Y}_n$ denote the sample means of x and y variables, and $\overline{y}_N$ be the population mean. Further, show that:

$$V\left(\hat{T}_r\right) = \frac{1}{\binom{N-1}{n-1}} \left[\sum_{t=1}^{\binom{N}{n}} \left\{ \frac{\left(\sum_{j=1}^{n} x_{i_j}\right)^2}{\sum_{j=1}^{n} p_{i_j}} \right\}_t \right] - T^2 \tag{5}$$

and an unbiased estimator of $V\left(\hat{T}_r\right)$ is given by

$$\hat{V}\left(\hat{T}_r\right) = \hat{R}^2 - \frac{1}{Nn\,\overline{Y}_n\,\overline{y}_N} \left\{ \sum_{j=1}^{n} x_{i_j}^2 + \frac{N-1}{n-1} \sum_{j=1}^{n} \sum_{\substack{j'=1 \\ j' \neq j}}^{n} x_{i_j} x_{i_{j'}} \right\} \tag{6}$$

[Lahiri (1951), Midzuno (1952), Sen (1952)]

4.4. Under the Midzuno scheme of sampling with replacement, consider the following estimator of population total T in T_3 - class

$$\hat{T}_3 = W_{i_1} \left[\sum_{r=1}^{n} (X_r = x_{i_r}) \right]$$

where, the weight W_{i_j} associated with the sample depends on the outcome of the first draw x_{i_1}. Show that $\hat{T}_3$ is MVLUE in T_3-class for $W_{i_1} = \frac{1}{n\,p_{i_1}}$. Also obtain expression of variance of $\hat{T}_3$.

4.5. Under the Midzuno scheme of sampling with replacement obtain unordered estimator $\hat{T}_{uo}$ corresponding to the following ordered estimator

$$\hat{T}_3 = \frac{1}{n\,p_{i_j}} \left[\sum_{j=1}^{k} x_{i_j} \,.\, R_{i_j} \right]$$

where, i_j-th unit is the unit occurring at the first draw and it is repeated R_{i_j} times in the given sample s_t.

Also obtain expressions of variance and an unbiased estimate of the variance of $\hat{T}_{uo}$.

4.6. Show that an unordered estiamtor $\hat{T}_{uo}$ corresponding to ordered estimator $\hat{T}_3$ given in Eq. (1) of Exercise 4.5, under Midzuno scheme of sampling with replacement is given by

$$\hat{T}_{uo} = \left(\sum_{j=1}^{k} x_{i_j} R_{i_j}\right) \Big/ \left(\sum_{j=1}^{k} R_{i_j} p_{i_j}\right)$$

where, k is number of distinct units in the sample $s_t \in S_p$.

Further, obtain combined unordered estiamtor $\hat{T}_{cu}$ corresponding to $\hat{T}_{uo}$ and give expression of variance of $\hat{T}_{cu}$.

CHAPTER 5

Estimation in T_4 - Class

5.1 INTRODUCTION

In T_4 –class of linear estimators, as given in Section 1.2 of Chapter 1, the weights associated with the observations on the units in the sample depend upon the units as well as on the draws at which these units occur in the sample. It will be shown in this chapter that this class is a non-empty class. We observe here in particular that Des Raj ordered estimator belongs to this class. Further, this estimator is not MVLUE as no such estimator exists in this class for sampling with or without replacement.

5.2 NON-EMPTINESS OF T_4 –CLASS

We prove that the T_4 –class of linear unbiased estimators is non-empty for sampling with or without replacement. For this, we note that a linear estimator in T_4 –class is

$$\hat{T}_4 = \sum_{r=1}^{n} \beta_{ro}^{st} X_r \tag{5.2.1}$$

where, $\beta_{ro}^{st} = \beta_{r\lambda}$ when $X_r = x_\lambda$ for r = 1, 2, . . ., n and $\lambda = 1, 2, \ldots, N$. The expected value of the estimator $\hat{T}_4$ is given by

$$E\left(\hat{T}_4\right) = \sum_{r=1}^{n} \sum_{\lambda=1}^{N} \beta_{r\lambda}\, p_{\lambda r}\, x_\lambda$$

$$= \sum_{\lambda=1}^{N} x_\lambda \left(\sum_{r=1}^{n} \beta_{r\lambda}\, p_{\lambda r} \right) \tag{5.2.2}$$

where, $p_{\lambda r}$ is the probability of λ–th unit occurring at r-th draw. For $\hat{T}_4$ to be unbiased estimator of the population total T, for all $X \in R^{(N)}$, it is necessary and sufficient that

$$\sum_{r=1}^{n} \beta_{r\lambda}\, p_{\lambda r} = 1 \tag{5.2.3}$$

for $\lambda = 1,2,\ldots,N$. By taking $\beta_{r\lambda} = 1/(n\, p_{\lambda r})$, we note that the condition of the consistency for the above system of Eqs. (5.2.3) is satisfied. Thus, T_4 – class is a non-empty class.

5.3 NON-EXISTENCE OF MVLUE IN THE CLASS FOR SWOR

For SWOR, the variance of $\hat{T}_4$ is given by

$$\text{Var}\left(\hat{T}_4\right) = E\left(\hat{T}_4^2\right) - T^2 \qquad (5.3.1)$$

where,

$$E\left(\hat{T}_4^2\right) = E\left(\sum_{r=1}^{n} \beta_{ro}\, X_r\right)^2$$

$$= E\left[\sum_{r=1}^{n} \beta_{ro}^2\, X_r^2 + \sum_{r=1}^{n} \sum_{s(\neq r)=1}^{n} \beta_{ro}\, \beta_{so}\, X_r X_s\right] \qquad (5.3.2)$$

$$= \sum_{\lambda=1}^{N} x_\lambda^2 \sum_{r=1}^{n} \beta_{r\lambda}^2\, p_{\lambda r} +$$

$$\sum_{\lambda=1}^{N} \sum_{\lambda'(\neq\lambda)=1}^{N} x_\lambda\, x_{\lambda'} \left\{\sum_{r=1}^{n} \sum_{s(\neq r)=1}^{n} \beta_{r\lambda}\, \beta_{s\lambda'}\, p(\lambda r, \lambda' s)\right\} \qquad (5.3.3)$$

where, $p(\lambda r, \lambda' s)$ is the probability of selecting the units λ and λ' at the r-th and s-th draws respectively for $r \neq s = 1,2,\ldots,n$ and $\lambda' \neq \lambda = 1, 2, \ldots, N$.

Let

$$\phi = E\left(\hat{T}_4^2\right) - T^2 - 2 \sum_{\lambda=1}^{N} \mu_\lambda \left[\sum_{r=1}^{n} \beta_{r\lambda}\, p_{\lambda r}\right]$$

where, μ_λ's are Lagrange's undetermined multipliers for the N conditions of unbiasedness given in Eq. (5.2.3). Now equating $\partial\phi/\partial\beta_{r\lambda} = 0$, we get

$$2\beta_{r\lambda}\, x_\lambda^2\, p_{\lambda r} + 2x_\lambda \sum_{s(\neq r)=1}^{n} \sum_{\lambda'(\neq\lambda)=1}^{N} \beta_{s\lambda'}\, x_{\lambda'}\, p(\lambda r, \lambda' s) - 2\mu_\lambda\, p_{\lambda r} = 0$$

or,

$$\beta_{r\lambda}\, x_\lambda^2\, p_{\lambda r} + x_\lambda \sum_{s(\neq r)=1}^{n} \sum_{\lambda'(\neq\lambda)=1}^{N} x_{\lambda'}\, \beta_{s\lambda'}\, p(\lambda r, \lambda' s) = \mu_\lambda\, p_{\lambda r}$$

If $p_{\lambda r} = 0$ for any r and λ, then it means that at that r-th draw λ-th unit does not occur in any sample and we do not have to bother about $\beta_{r\lambda}$. Therefore, we need to consider those $\beta_{r\lambda}$'s for which $p_{\lambda r} \neq 0$. Further we note that

$$p_{\lambda r} = \sum_{\lambda'(\neq\lambda)=1}^{N} \sum_{s(\neq r)=1}^{n} p(\lambda r, \lambda' s)$$

When $p_{\lambda r} \neq 0$, there exists some λ' and s such that $P(\lambda r, \lambda' s) \neq 0$. Now, consider the population vector X in which $x_\lambda = 1$ and $x_\nu = 0$ for all $\nu \neq \lambda$. Then, $\beta_{r\lambda}\, p_{\lambda r} = \mu_\lambda\, p_{\lambda r}$ giving

$$\beta_{r\lambda} = \mu_\lambda \qquad \text{for all r and } \lambda \tag{5.3.4}$$

Further consider the population vector X where $x_\lambda = x_{\lambda'} = 1$ and $x_\nu = 0$ for all $\nu \neq \lambda, \lambda'$. Then,

$$\beta_{r\lambda}\, p_{\lambda r} + \sum_{s(\neq r)=1}^{n} \beta_{s\lambda'}\, p(\lambda r, \lambda' s) = \mu_\lambda\, p_{\lambda r}.$$

Since $\beta_{r\lambda} = \mu_\lambda$ for all r and λ,

$$\mu_\lambda\, p_{\lambda r} + \mu_\lambda \sum_{s(\neq r)=1}^{n} p(\lambda r, \lambda' s) = \mu_\lambda\, p_{\lambda r}$$

or,

$$\mu_\lambda \sum_{s(\neq r)=1}^{n} p(\lambda r, \lambda' s) = 0$$

Since at least one $p(\lambda r, \lambda' s) \neq 0$, $\mu_\lambda = 0$. This gives $\beta_{r\lambda} = 0$, for all r and λ. Thus, there is a contradiction to the conditions of unbiasedness for $\hat{T}_4$ in Eq. (5.2.3). Hence the MVLUE in T_4-class of estimators does not exist in case of SWOR.

5.4 NON-EXISTENCE OF MVLUE IN THE CLASS FOR SWR

For sampling with replacement,

$$V\left(\hat{T}_4\right) = \sum_{r=1}^{n} \text{Var}\,(\beta_{ro}\, X_r) \tag{5.4.1}$$

In order to determine the optimum values of the weights β's, let

$$\phi = \sum_{r=1}^{n} \text{Var}(\beta_{ro} X_r) - 2 \sum_{\lambda=1}^{N} \mu_\lambda \sum_{r=1}^{n} \beta_{r\lambda} p_{\lambda r}$$

$$= \sum_{r=1}^{n} \left[E(\beta_{ro} X_r)^2\right] - \sum_{r=1}^{n} \left[\sum_{\lambda=1}^{N} \beta_{r\lambda} x_\lambda p_{\lambda r}\right]^2$$

$$- 2 \sum_{\lambda=1}^{N} \mu_\lambda \left(\sum_{r=1}^{n} \beta_{r\lambda} p_{\lambda r}\right)$$

$$= \sum_{r=1}^{n} \sum_{\lambda=1}^{N} (\beta_{r\lambda} x_\lambda)^2 p_{\lambda r} - \sum_{r=1}^{n} \left(\sum_{\lambda=1}^{N} \beta_{r\lambda} x_\lambda p_{\lambda r}\right)^2$$

$$- 2 \sum_{\lambda=1}^{N} \mu_\lambda \sum_{r=1}^{n} \beta_{r\lambda} p_{\lambda r}$$

where, $\mu_1, \mu_2, \ldots, \mu_N$ are N Lagrange's multipliers for the N conditions of unbiasedness given in Eq. (5.2.3). Differentiating ϕ w.r.t. $\beta_{r\lambda}$ and equating the differential coefficient to zero we get,

$$\frac{\partial \phi}{\partial \beta_{r\lambda}} = 2 \beta_{r\lambda} x_\lambda^2 p_{\lambda r} - 2\left(\sum_{\lambda=1}^{N} \beta_{r\lambda} x_\lambda p_{\lambda r}\right) p_{\lambda r} x_\lambda - 2 \mu_\lambda p_{\lambda r} = 0$$

or,

$$x_\lambda^2 \beta_{r\lambda} - x_\lambda \left(\sum_{\lambda=1}^{N} \beta_{r\lambda} x_\lambda p_{\lambda r}\right) = \mu_\lambda \tag{5.4.2}$$

by considering only those $\beta_{r\lambda}$'s for which $p_{\lambda r} \neq 0$. Now let us consider the population vector X where $x_\lambda = 1$ and $x_\nu = 0$ for all $\nu \neq \lambda$. Then, from Eq. (5.4.2) we get

$$\beta_{r\lambda} - \beta_{r\lambda} p_{\lambda r} = \mu_\lambda \tag{5.4.3}$$

Let us now consider the population vector X with $x_\lambda = x_{\lambda'} = 1$ and $x_\nu = 0$ for $\nu \neq \lambda, \lambda'$. Then from Eq. (5.4.2) we get

$$\beta_{r\lambda} - \beta_{r\lambda} p_{\lambda r} - \beta_{r\lambda'} p_{\lambda' r} = \mu_\lambda$$

Using Eq. (5.4.3),

$$\beta_{r\lambda'} p_{\lambda' r} = 0 \tag{5.4.4}$$

Since $p_{\lambda' r}$'s are not taken equal to zero; $\beta_{r\lambda'} = 0$ for all r and λ'. This then contradicts our conditions of unbiasedness given in Eq. (5.2.3). Thus we conclude that MVLUE in T_4-class, independent of population values, does not exist even in the case of SWR.

5.5 DES RAJ ORDERED ESTIMATOR

Let p_i be the probability of drawing i-th unit at the first draw for i= 1, 2, . . ., N. Thus, $\sum_{i=1}^{N} (p_i) = 1$. Let $i_1, i_2, \ldots, i_{r-1}$ be the units drawn in the first (r–1) draws. Then probability of i_r-th unit being drawn at r-th draw, under proposed sampling scheme without replacement, is $p_{i_r} / \left(1 - \sum_{j=1}^{r-1} p_{i_j}\right)$. This is the sampling scheme considered earlier by Horvitz and Thompson (1952). The ordered estimator of population total (Des Raj, 1956) under this sampling scheme without replacement, is

$$\hat{T} = \sum_{r=1}^{n} Z_r \tag{5.5.1}$$

where,

$$Z_r = \left[\sum_{j=1}^{r-1} (x_{i_j}) + x_{i_r} \frac{\left(1 - \sum_{j=1}^{r-1} p_{i_j}\right)}{p_{i_r}}\right] \tag{5.5.2}$$

The expression of the variance of $\hat{T}$ is complicated. However, an unbiased estimator of this variance is

$$\frac{1}{n(n-1)} \sum_{r=1}^{n} (Z_r - \overline{Z})^2 \tag{5.5.3}$$

where, $\overline{Z} = (\sum Z_r)/n$.

That $\hat{T}$ belongs to T_4-class is seen by noting that the coefficient of x_{i_r}, for $r = 1, 2, \ldots, n$ in $\hat{T}$ is

$$\frac{\left(1 - \sum_{j=1}^{r-1} p_{i_j}\right)}{p_{i_r}} + (n-r)$$

$$= \frac{1}{p_{i_r}} \left[1 - \sum_{j=1}^{r-1} p_{i_j} + (n-r)\, p_{i_r}\right]$$

which depends on the unit drawn at the r-th draw as well as on the draw itself. Since MVLUE does not exist in T_4-class, this estimator is not MVLUE. But it has been shown by Des Raj (1956) that this estimator has smaller variance than the variance of $\hat{T}_{2w}$, for sampling with replacement with $p_{i_r} = p_i$, given in Section 3.4 of Chapter 3. Using the results of Section 1.4 of Chapter 1, this estimator can be further unordered and therefore improved upon. However, combined unordering is not possible, as this estimator is for sampling without replacement scheme.

EXERCISES

5.1. Define T_4 -class of linear estimators and show that MVLUE does not exist in this class for varying probability of sampling without replacement.

5.2. Show that under Midzuno scheme of sampling without replacement, the MVLUE does not exist in T_4 -class.

5.3. Show that under Horvitz-Thompson sampling scheme (for n=2), the MVLUE exists in T_4 -class.

5.4. For sampling with varying probabilities with or without replacement, obtain unordered estimator $\hat{T}_{uo}$ corresponding to ordered estimator

$$\hat{T}_o = \sum_{r=1}^{n} \beta_{ro} X_r$$

of population total T in T_4 -class. Further, obtain expressions of variance and unbiased estimate of the variance of $\hat{T}_{uo}$.

5.5 Discuss the technique of combined unordering for sampling with varying probabilities with replacement and obtain a combined unordered estimator $\hat{T}_{cu}$ corresponding to unordered estimator $\hat{T}_{uo}$, obtained in Exercise 5.4. Also obtain expressions of variance and unbiased estimate of the variance of $\hat{T}_{cu}$.

[Bhargava, N.K. (1975)]

CHAPTER 6

Estimation in T_5, T_6 and T_7 - Classes

6.1 INTRODUCTION

In this chapter, we discuss linear estimation in the remaining T-classes, T_5, T_6 and T_7, taking them one by one in order. We give here a unified proof for the non-existence of MVLUE in T_5-class thereby showing that Basu's estimator (1958) for SRSWR scheme, which belongs to T_5-class (See Section 1.9, Chapter 1) is not MVLUE. We further prove non-existence of MVLUE in T_7-class which is more general than the general class due to Godambe (1955). Many estimators of practical interest, arising out to Godambe's general class, are presented in Section 6.7. Earlier in Sections 6.4 and 6.5, we prove that the classical estimators for SRSWOR and SRSWR schemes are MVLUE in T_6-class also. Now we first present results for T_5-class.

6.2 ESTIMATION IN T_5 – CLASS FOR SAMPLING WITH OR WITHOUT REPLACEMENT

Let a given s_t consist of $(i_1, i_2, \ldots, i_k)$ k distinct units with i_j-th unit occurring $R_{i_j}^{(p)}$ times for j = 1, 2, . . ., k in the sample such that $\sum_{j=1}^{k} R_{i_j}^{(p)} = n$. Then from Section 1.2 of Chapter 1, an estimator, based on sample s_t in T_5-class is

$$\hat{T}_5 = \left[\sum_{j=1}^{k}\left(\beta_{i_j}^{S_p}\ R_{i_j}^{(p)}\ x_{i_j}\right)\right]_{s_t} \tag{6.2.1}$$

Thus the total number of weights $\beta_i^{S_p}$ is

$$\left[\sum_{k=1}^{n} k\left\{\sum_{l=1}^{\binom{N}{k}} (M_{kl})\right\}\right] = \sum_{k=1}^{n}\left[\binom{N}{k}\binom{n-1}{k-1}k\right] \tag{6.2.2}$$

We determine these weights $\beta_i^{S_p}$ such that $E(\hat{T}_5) = T$ for all $X \in R^{(N)}$. We further examine whether T_5-class is non-empty for such values of the weights which make the estimator unbiased. We then prove the non-existence of MVLUE by minimising $V(\hat{T}_5)$ subject to the conditions of unbiasedness, for all $X \in R^{(N)}$. Let S' be the sample space, as in Chapter 4. Then

$$E(\hat{T}_5) = \sum_{s_t \in S'} \left[\sum_{j=1}^{k} R_{i_j}^{(p)} \beta_{i_j}^{S_p} x_{i_j}\right]_{s_t} p(s_t)$$

$$= \sum_{k=1}^{n} \sum_{l=1}^{\binom{N}{k}} \sum_{p=1}^{M_{kl}} \sum_{t=1}^{M_{klp}} \left[\sum_{j=1}^{k} R_{i_j}^{(p)} \beta_{i_j}^{S_p} x_{i_j}\right]_{s_t} p(s_t)$$

$$= \sum_{k=1}^{n} \sum_{l=1}^{\binom{N}{k}} \sum_{p=1}^{M_{kl}} \left[\sum_{j=1}^{k} R_{i_j}^{(p)} \beta_{i_j}^{S_p} x_{i_j}\right] p(S_p)$$

where, $$p(S_p) = \sum_{s_t \in S_p} p(s_t) = \sum_{t=1}^{M_{klp}} p(s_t).$$

Therefore,

$$E(\hat{T}_5) = \sum_{i=1}^{N} x_i \left[\sum_{k=1}^{n} \sum_{l=1}^{\binom{N-1}{k-1}} \sum_{p=1}^{M_{kl}} R_i^{(p)} \beta_i^{S_p} p(S_p)\right]$$

In order that $\hat{T}_5$ be unbiased for the population total T (X) for all $X \in R^{(N)}$,

$$\sum_{k=1}^{n} \sum_{l=1}^{\binom{N-1}{k-1}} \sum_{p=1}^{M_{kl}} p(S_p) R_i^{(p)} \beta_i^{S_p} = 1 \tag{6.2.3}$$

for, $i = 1, 2, \ldots, N.$

Let $$\beta_i^{S_p} = 1/\left(p(S_p)\, C\, R_i^{(p)}\right) \tag{6.2.4}$$

for all $i \in s_t \in S_p$ and for all S_p with

$$C = \left[\sum_{k=1}^{n} \sum_{l=1}^{\binom{N-1}{k-1}} \sum_{p=1}^{M_{kl}} (1)\right] \quad \text{for } i = 1,2,\ldots, N.$$

This set of $\beta_i^{S_p}$ satisfies the above N eqautions arising out of the conditions of unbiasedness. Hence, the N equations are consistent.

We now proceed to prove non-existence of MVLUE in the class. The variance of the estimator $\hat{T}_5$ is given by

$$V\left(\hat{T}_5\right) = E\left(\hat{T}_5^2\right) - T^2$$

$$= \sum_{s_t \in S'} \left[\sum_{j=1}^{k} R_{i_j}^{(p)} \beta_{i_j}^{S_p} x_{i_j}\right]_{s_t}^2 p(s_t) - T^2$$

$$= \sum_{k=1}^{n} \sum_{l=1}^{\binom{N}{k}} \sum_{p=1}^{M_{kl}} \left[\sum_{j=1}^{k} R_{i_j}^{(p)} \beta_{i_j}^{S_p} x_{i_j}\right]^2 p(S_p) - T^2 \tag{6.2.5}$$

In order to get the optimum values of $\beta_i^{S_p}$, let

$$\phi = V\left(\hat{T}_5\right) - 2\sum_{i=1}^{N} \lambda_i \left[\sum_{k=1}^{n} \sum_{l=1}^{\binom{N-1}{k-1}} \sum_{p=1}^{M_{kl}} R_i^{(p)} \beta_i^{S_p} p(S_p) - 1\right]$$

$$= \sum_{k=1}^{n} \sum_{l=1}^{\binom{N}{k}} \sum_{p=1}^{M_{kl}} \left[\sum_{j=1}^{k} R_{i_j}^{(p)} x_{i_j} \beta_{i_j}^{S_p}\right]^2 p(S_p) - T^2$$

$$- 2\sum_{i=1}^{N} \lambda_i \left[\sum_{k=1}^{n} \sum_{l=1}^{\binom{N-1}{k-1}} \sum_{p=1}^{M_{kl}} R_i^{(p)} \beta_i^{S_p} p(S_p) - 1\right] \tag{6.2.6}$$

where, λ_i is Lagrange's undetermined multiplier corresponding to the i-th Eq. (6.2.3) for i = 1, 2, . . ., N. Equating $\partial\phi / \partial\beta_i^{S_p}$ to zero, we get for an s_t containing i-th unit and $\in S_p$

$$\left(\sum_{j=1}^{k} R_{i_j}^{(p)} \beta_{i_j}^{S_p} x_{i_j}\right)_{s_t} x_i = \lambda_i \tag{6.2.7}$$

The choice of $\beta_i^{S_p}$ given by Eq. (6.2.7) should hold for all $X \in R^{(N)}$. We consider a particular population vector $X = [0, 0, \ldots, x_i (\neq 0), 0, 0, \ldots, 0]$, then Eq. (6.2.7) implies that $R_i^{(p)} \beta_i^{S_p} x_i^2 = \lambda_i$. For population vector X with all x_i lying in the given

s_t being zero; then it implies that $\lambda_i = 0$ which in return gives $\beta_i^{S_p} = 0$. Since this is true for arbitrary S_p and i, we have $\beta_i^{S_p} = 0$ for all i and S_p. However, the set of $\beta_i^{S_p}$ with all their values zero, does not satisfy the conditions of unbiasedness, thereby bringing a contradiction. Thus MVLUE does not exist in the class.

6.3 ESTIMATION IN T_6 – CLASS FOR SAMPLING WITH OR WITHOUT REPLACEMENT

Let a given sample $s_t \in S_p\left[\left(i_1, R_{i_1}^{(p)}\right), \ldots, \left(i_k, R_{i_k}^{(p)}\right)\right]$. Then, from Section 1.2 of Chapter 1, an estimator in T_6-class based on s_t is

$$\hat{T}_6 = \left[\sum_{r=1}^{n} \left(\beta_r^{S_p} X_r\right)\right]_{s_t} \tag{6.3.1}$$

The total number of weights $\beta_r^{S_p}$ is $n\left[\sum_{k=1}^{n} \binom{N}{k}\binom{n-1}{k-1}\right]$. We determine these weights $\beta_r^{S_p}$ such that $E\left(\hat{T}_6\right) = T$ for all $X \in R^{(N)}$. Further we find the MVLUE for all $X \in R^{(N)}$ if it exists, by minimising the variance of $\hat{T}_6$ subject to the conditions of unbiasedness.

$$E\left(\hat{T}_6\right) = \sum_{s_t \in S'} \left(\sum_{r=1}^{n} \beta_r^{S_p} X_r\right)_{s_t} p(s_t)$$

$$= \sum_{k=1}^{n} \sum_{l=1}^{\binom{N}{k}} \sum_{p=1}^{M_{kl}} \sum_{t=1}^{M_{klp}} \left[\left(\sum_{r=1}^{n} \beta_r^{S_p} X_r\right)_{s_t} p(s_t)\right]$$

$$= \sum_{i=1}^{N} x_i \left[\sum_{k=1}^{n} \sum_{l=1}^{\binom{N-1}{k-1}} \sum_{p=1}^{M_{kl}} \sum_{t=1}^{M_{klp}} \left[\left(\sum_{r=1}^{n} \beta_r^{Sp} \delta_r\right)_{s_t} p(s_t)\right]\right]$$

where,

$$\delta_{ri} = \begin{cases} 1, & \text{if i-th unit occurs at r-th draw in a given } s_t \\ 0, & \text{otherwise.} \end{cases}$$

In order that $\hat{T}_6$ be an unbiased estimator for the population total T for all $X \in R^{(N)}$, we must have

$$\sum_{k=1}^{n} \sum_{l=1}^{\binom{N-1}{k-1}} \sum_{p=1}^{M_{kl}} \sum_{t=1}^{M_{klp}} \left[\left(\sum_{r=1}^{n} \beta_r^{S_p} \delta_{ri} \right)_{s_t} p(s_t) \right] = 1$$

or,

$$\sum_{k=1}^{n} \sum_{l=1}^{\binom{N-1}{k-1}} \sum_{p=1}^{M_{kl}} \sum_{r=1}^{n} \left[\beta_r^{S_p} \sum_{t=1}^{M_{klp}} \{ p(s_t) \delta_{ri} \}_{s_t} \right] = 1$$

for $i = 1, 2, \ldots, N$. (6.3.2)

In order to prove that these N equations are consistent, it is enough to show that there exists at least one set of $\beta_r^{S_p}$'s which satisfy all the equations. For this, let

$$\beta_r^{S_p} = 1 / \left[C.n \sum_{t=1}^{M_{klp}} \{ p(s_t) \delta_{ri} \}_{s_t} \right] \quad (6.3.3)$$

for all $r = 1, 2, \ldots, n$ and all $S_p \in S'$, where C is given by

$$C = \left[\sum_{k=1}^{n} \sum_{l=1}^{\binom{N-1}{k-1}} \sum_{p=1}^{M_{kl}} (1) \right] = \left[\sum_{k=1}^{n} \binom{N-1}{k-1} \binom{n-1}{k-1} \right].$$

It may be noted that for sampling without replacement, the expression for C reduces to $\binom{N-1}{n-1}$.

It appears from Eq. (6.3.3) that a particular $\beta_r^{S_p}$ may depend upon the occurrence of a particular unit at r-th draw in different samples $s_t \in S_p$. That, it does not for SRSWOR and SRSWR, is shown in the following Sections 6.4 and 6.5. In such cases T_6-class is non-empty. However, there are situations where the weights do depend upon the occurrence of a particular unit. In situations of the kind, the estimators lie in an entirely new class of linear estimators due to Tikkiwal, G.C. (1988) and T_6-class as such is empty. This new class depends upon both the draw and the outcome of the unit at that draw in a given S_p.

The variance of an estimator in T_6-class is given by

$$V(\hat{T}_6) = E(\hat{T}_6^2) - T^2$$

$$= \sum_{s_t \in S'} \left[\left(\sum_{r=1}^{n} \beta_r^{S_p} X_r \right)_{s_t}^2 p(s_t) \right] - T^2$$

$$= \sum_{k=1}^{n} \sum_{l=1}^{\binom{N}{k}} \sum_{p=1}^{M_{kl}} \sum_{t=1}^{M_{klp}} \left[\left(\sum_{r=1}^{n} \beta_r^{S_p} X_r \right)_{s_t}^2 p(s_t) \right] - T^2 \tag{6.3.4}$$

Let

$$\phi = V(\hat{T}_6) - 2 \sum_{i=1}^{N} \lambda_i \left[\sum_{k=1}^{n} \sum_{l=1}^{\binom{N-1}{k-1}} \sum_{p=1}^{M_{kl}} \sum_{r=1}^{n} \beta_r^{S_p} \left\{ \sum_{t=1}^{M_{klp}} p(s_t) \delta_{ri} \right\}_{s_t} - 1 \right]$$

$$= \sum_{k=1}^{n} \sum_{l=1}^{\binom{N}{k}} \sum_{p=1}^{M_{kl}} \sum_{t=1}^{M_{klp}} \left[\left(\sum_{r=1}^{n} \beta_r^{S_p} X_r \right)_{s_t}^2 p(s_t) \right] - T^2$$

$$- 2 \sum_{i=1}^{N} \lambda_i \left[\sum_{k=1}^{n} \sum_{l=1}^{\binom{N-1}{k-1}} \sum_{p=1}^{M_{kl}} \sum_{r=1}^{n} \beta_r^{S_p} \sum_{t=1}^{M_{klp}} \{ p(s_t) \delta_{ri} \}_{s_t} - 1 \right] \tag{6.3.5}$$

where, λ_i is Lagrange's undetermined multiplier corresponding to the Eq. (6.3.2) for i = 1, 2, . . . , N. Equating $\partial\phi / \partial\beta_r^{S_p}$ equal to zero, we get for an s_t containing i-th unit and $\in S_p$

$$\sum_{t=1}^{M_{klp}} \left[\left(\sum_{r=1}^{n} \beta_r^{S_p} X_r \right)_{s_t} x_i \, p(s_t) \right]$$

$$= \sum_{i \in s_t \in S_p} \lambda_i \left[\sum_{t=1}^{M_{klp}} \{ p(s_t) \delta_{ri} \}_{s_t} \right] \tag{6.3.6}$$

for different $\beta_r^{S_p}$.

The MVLUE will now be determined, in case of a given design, by solving Eqs. (6.3.2) and (6.3.6). We now proceed to obtain MVLUE in special cases.

6.4 MVLUE IN T_6 -CLASS FOR SRSWOR

When there is simple random sampling without replacement,

$$\sum_{t=1}^{M_{klp}} [p(s_t) \delta_{ri}] = (n-1)! \Big/ \left\{ n! \binom{N}{n} \right\}.$$

Therefore, in this case $\beta_r^{S_p} = N/n$ for all r and S_p from Eq. (6.3.3), as $C = \binom{N-1}{n-1}$ for sampling without replacement.

Thus, the equations of unbiasedness are consistent and T_6-class is non-empty for this sampling design. Further, since $\beta_r^{S_p} = N/n$ is one of the solutions, the classical estimator lies in this class.

Now under this sampling scheme, Eq. (6.3.6) reduces to

$$\sum_{t=1}^{M_{klp}} \left[\left(\sum_{r=1}^{n} \beta_r^{S_p} X_r \right)_{s_t} (x_{i_r}) \right] = \sum_{i \in s_t \in S_p} \lambda_i \sum_{t=1}^{M_{klp}} (\delta_{ri})_{s_t}$$

$$= (n-1)! \left[\sum_{i \in s_t \in S_p} \lambda_i \right] \quad \text{for different } S_p \text{ and r.}$$

For each S_p, we are having a set of n such equations for $r = 1, 2, \ldots, n$. The right hand side is same for all these equations.

Taking any two equations one for r and other for $r' (\neq r)$ for a given S_p, we have

$$\sum_{t=1}^{M_{klp}} \left(\sum_{r=1}^{n} \beta_r^{S_p} X_r \right)_{s_t} x_{i_r} = \sum_{t=1}^{M_{klp}} \left(\sum_{r=1}^{n} \beta_r^{S_p} X_r \right)_{s_t} x_{i_{r'}}$$

with x_{i_r} and $x_{i_{r'}}$ denoting the variate values of two different units occurring at r and r′ draws in s_t for $t = 1, 2, \ldots, M_{klp}$ samples in S_p;

or,

$$\beta_r^{S_p} \left[\sum_{t=1}^{M_{klp}} (x_{i_r}^2)_{s_t} - \sum_{t=1}^{M_{klp}} (x_{i_r} x_{i_{r'}})_{s_t} \right]$$

$$- \beta_{r'}^{S_p} \left[\sum_{t=1}^{M_{klp}} (x_{i_{r'}}^2)_{s_t} - \sum_{t=1}^{M_{klp}} (x_{i_r} x_{i_{r'}})_{s_t} \right]$$

$$+ \sum_{r''(\neq r, r')}^{n} \beta_{r''}^{S_p} \left[\sum_{t=1}^{M_{klp}} (x_{i_r} x_{i_{r''}})_{s_t} - \sum_{t=1}^{M_{klp}} (x_{i_{r'}} x_{i_{r''}})_{s_t} \right] = 0$$

Since,

$$\sum_{t=1}^{M_{kl/p}} (x_{i_s}^2)_{s_t} = (n-1)!\left[\sum_{i_j \in S_p} x_{i_j}^2\right], \qquad \text{for } s = r, r'$$

and,

$$\sum_{t=1}^{M_{kl/p}} (x_{i_s}\, x_{i_{s'}})_{s_t} = (n-2)!\left[\sum_{i_j,\, i_k \in S_p (j \neq k)} x_{i_j}\, x_{i_k}\right]$$

for any two s and s′ draws;

We have

$$\left[\beta_r^{S_p} - \beta_{r'}^{S_p}\right]\left[\sum_{t=1}^{M_{kl/p}} (x_{i_r}^2)_{s_t} - \sum_{t=1}^{M_{kl/p}} (x_{i_r}\, x_{i_{r'}})_{s_t}\right] = 0$$

Since, in general, the second factor is not equal to zero $[\beta_r^{S_p} - \beta_{r'}^{S_p}] = 0$. As it is true for any arbitrary r, r′ and S_p; $\beta_r^{S_p} = K$ for all S_p and r. That K = N/n is now seen from Eq. (6.3.2). Thus the classical estimator is itself MVLUE in this class for SRSWOR.

6.5 MVLUE IN T_6 – CLASS FOR SRSWR

In order to prove that T_6-class is non-empty for simple random sampling with replacement, we note from Eq. (6.3.3) that $\beta_r^{S_p} = C'$, a constant quantity is a solution to the set of Eq. (6.3.2) for different i. Thus, from Eq. (6.3.2), we have

$$\frac{C'}{N^n}\left[\sum_{k=1}^{n} \sum_{l=1}^{\binom{N-1}{k-1}} \sum_{p=1}^{M_{kl}} \sum_{r=1}^{n} \sum_{t=1}^{M_{kl/p}} (\delta_{ri})_{s_t}\right] = 1$$

As the expression within bracket gives the number of times i-th unit occurs in different N^n samples, from Eq. (4.5.4) of Chapter 4 we have

$$C' = \frac{N^n}{n\, N^{n-1}} = \frac{N}{n}$$

Thus,

$$\beta_r^{S_p} = \frac{N}{n}, \qquad \text{for all r and } S_p \tag{6.5.1}$$

and so the classical estimator for SRSWR lies in T_6-class. In order to show that this estimator is itself MVLUE in this class, we proceed as follows.

Under this sampling scheme Eq. (6.3.6) reduces to

$$\left[\sum_{t=1}^{M_{k/p}} \left(\sum_{r=1}^{n} \beta_r^{S_p} X_r\right)_{s_t} (x_{i_r})_{s_t}\right] = \sum_{i \in s_t \in S_p} \lambda_i \sum_{t=1}^{M_{k/p}} (\delta_{ri})_{s_t} \tag{6.5.2}$$

For each S_p, we have a set of n such equations for r = 1, 2, . . ., n. The RHS is same for all these equations.

Thus, taking any two equations one for r and other for $r' (\neq r)$ for a given S_p, we have

$$\left[\sum_{t=1}^{M_{k/p}} \left(\sum_{r=1}^{n} \beta_r^{S_p} X_r\right)_{s_t} (x_{i_r})_{s_t}\right] = \left[\sum_{t=1}^{M_{k/p}} \left(\sum_{r=1}^{n} \beta_r^{S_p} X_r\right)_{s_t} (x_{i_{r'}})_{s_t}\right]$$

or,

$$\beta_r^{S_p}\left[\sum_{t=1}^{M_{k/p}} (x_{i_r}^2)_{s_t} - \sum_{t=1}^{M_{k/p}} (x_{i_r} x_{i_{r'}})_{s_t}\right]$$

$$-\beta_{r'}^{S_p}\left[\sum_{t=1}^{M_{k/p}} (x_{i_{r'}}^2)_{s_t} - \sum_{t=1}^{M_{k/p}} (x_{i_r} x_{i_{r'}})_{s_t}\right]$$

$$+\sum_{r''(\neq r, r')}^{n} \beta_{r''}^{S_p}\left[\sum_{t=1}^{M_{k/p}} (x_{i_r} x_{i_{r''}})_{s_t} - \sum_{t=1}^{M_{k/p}} (x_{i_{r'}} x_{i_{r''}})_{s_t}\right] = 0 \tag{6.5.3}$$

Since the last expression vanishes as in Section 6.4,

$$\left[\beta_r^{S_p} - \beta_{r'}^{S_p}\right]\left[\sum_{t=1}^{M_{k/p}} (x_{i_r}^2)_{s_t} - \sum_{t=1}^{M_{k/p}} (x_{i_r} x_{i_{r'}})_{s_t}\right] = 0 \tag{6.5.4}$$

Since, in general, the second factor of Eq. (6.5.4) is not equal to zero, $\left[\beta_r^{S_p} - \beta_{r'}^{S_p}\right] = 0$. As it is true for any arbitrary r, r′ and S_p; $\beta_r^{S_p} = K(\text{constant})$ for all S_p and r. That K = N/n follows from what is said in the beginning of this Section. Thus the classical estimator is MVLUE in this T_6 – class for SRSWR as asserted.

6.6 ESTIMATION IN T_7 – CLASS

From Section 1.2 of Chapter 1, an estimator based on sample s_t in T_7 -class is

$$\hat{T}_7 = \sum_{r=1}^{n} \beta_{ro}^{s_t} X_r \qquad (6.6.1)$$

where, $\beta_{ro}^{s_t} = \beta_{ri}^{s_t}$, if $X_r = x_i$ for all r and i in s_t. Thus the total number of weights $\beta_{ro}^{s_t}$ is n. N^n which are fixed in advance.

We determine these weights $\beta_{ro}^{s_t}$ such that $E(\hat{T}_7) = T$ for all $X \in R^{(N)}$. Further we find the MVLUE, if it exists, by minimising $V(\hat{T}_7)$ for all $X \in R^{(N)}$ subject to the conditions of unbiasedness.

$$E(\hat{T}_7) = \sum_{s_t \in S'} \left[\sum_{r=1}^{n} \beta_{ro}^{s_t} X_r \right] p(s_t)$$

$$= \sum_{k=1}^{n} \sum_{l=1}^{\binom{N}{k}} \sum_{p=1}^{M_{kl}} \sum_{t=1}^{M_{klp}} \left[\sum_{r=1}^{n} \beta_{ro}^{s_t} X_r \right] p(s_t)$$

$$= \sum_{i=1}^{N} x_i \left[\sum_{k=1}^{n} \sum_{l=1}^{\binom{N-1}{k-1}} \sum_{p=1}^{M_{kl}} \sum_{t=1}^{M_{klp}} \left\{ \sum_{r=1}^{n} \beta_{ro}^{s_t} \delta_{ri} \right\} p(s_t) \right]$$

where, $\delta_{ri} = 1$, if i-th unit occurs at the r-th draw in a given $s_t \in S'$ and equal to zero, otherwise.

So, for $\hat{T}_7$ to be unbiased for the population total T for all $X \in R^{(N)}$, we have

$$\sum_{k=1}^{n} \sum_{l=1}^{\binom{N-1}{k-1}} \sum_{p=1}^{M_{kl}} \sum_{t=1}^{M_{klp}} \left[\left\{ \sum_{r=1}^{n} \beta_{ro}^{s_t} \delta_{ri} \right\} p(s_t) \right] = 1 \qquad (6.6.2)$$

for $i = 1, 2, \ldots, N$.

For $n > 1$, the number of unknowns $\beta_{ro}^{s_t}$ will be more than N so that the number of solutions will be infinite. The choice of

$\beta_{ro}^{s_t} = 1/\{p(s_t)\,Cn\}$ with C equal to $\left[\sum_{k=1}^{n}\sum_{l=1}^{\binom{N-1}{k-1}}\sum_{p=1}^{M_{kl}} M_{klp}\right]$ satisfies the system of N equations given by Eq. (6.6.2). Hence T_7-class is non-empty.

The variance of an estimator in T_7-class is given by

$$V\left(\hat{T}_7\right) = E\left(\hat{T}_7^2\right) - T^2$$

$$= \sum_{s_t \in S'} \left[\left(\sum_{r=1}^{n} \beta_{ro}^{s_t} X_r\right)^2 p(s_t)\right] - T^2$$

$$= \sum_{k=1}^{n}\sum_{l=1}^{\binom{N}{k}}\sum_{p=1}^{M_{kl}}\sum_{t=1}^{M_{klp}} \left[\left(\sum_{r=1}^{n} \beta_{ro}^{s_t} X_r\right)^2 p(s_t)\right] - T^2 \qquad (6.6.3)$$

Let,

$$\phi = V\left(\hat{T}_7\right) - 2\sum_{i=1}^{N} \lambda_i \left[\sum_{k=1}^{n}\sum_{l=1}^{\binom{N-1}{k-1}}\sum_{p=1}^{M_{kl}}\sum_{t=1}^{M_{klp}} \left\{\sum_{r=1}^{n} \beta_{ro}^{s_t} \delta_{ri}\right\} p(s_t) - 1\right]$$

where, λ_i is Lagrange's multiplier corresponding to the i-th equation of Eq. (6.6.2) for i = 1, 2, . . ., N.

$$= \sum_{k=1}^{n}\sum_{l=1}^{\binom{N}{k}}\sum_{p=1}^{M_{kl}}\sum_{t=1}^{M_{klp}} \left[\sum_{r=1}^{n} \left(\beta_{ro}^{s_t} X_r\right)^2 p(s_t)\right] - T^2$$

$$- 2\sum_{i=1}^{N} \lambda_i \left[\sum_{k=1}^{n}\sum_{l=1}^{\binom{N-1}{k-1}}\sum_{p=1}^{M_{kl}}\sum_{t=1}^{M_{klp}} \left\{\sum_{r=1}^{n} \beta_{ro}^{s_t} \delta_{ri}\right\} p(s_t) - 1\right] \qquad (6.6.4)$$

For any given $\beta_{ro}^{s_t}$, equating $\partial\phi/\partial\beta_{ro}^{s_t}$ equal to zero, we get

$$\left[\left(\sum_{r=1}^{n} \beta_{ro}^{s_t} X_r\right) x_{i_r}\right] = \lambda_{i_r} \qquad (6.6.5)$$

In order that MVLUE exist in this class, the choice of $\beta_{ro}^{s_t}$ given by Eq. (6.6.5) should be independent of all $X \in R^{(N)}$. Let us consider the population vector $X = [\,x_{i_r} = 1;\, x_{i'} = 0$ for $i'(\neq i_r) = 1, 2, \ldots, N]$. Then

Eq. (6.6.5) gives $\beta_{ro}^{s_t} = \lambda_{i_r}$. For the population vector $X = [x_i = 0$ for all $i \in s_t$ and arbitrary values for others], the Eq. (6.6.5) gives $\lambda_{i_r} = 0$. Thus, $\beta_{ro}^{s_t} = 0$ for all s_t, r and i. But this choice of $\beta_{ro}^{s_t}$'s does not satisfy the conditions of unbiasedness given by Eq. (6.6.2), thus leading to a contradiction. Hence MVLUE does not exist in this class.

6.7 SOME RESULTS DUE TO GODAMBE

Godambe (1955, section 2, p. 269) gave a general class of linear estimators, which is shown to be a special case of T_7-class in Theorem 1.3.1 of Chapter 1. He showed that MVLUE does not exist in his general class. A compact and more elegant proof of this can be easily obtained from what is given in Section 6.6. Therefore, we omit this and proceed to discuss other results of Godambe obtained by him in the process of obtaining MVLUE estimator in the general class. His general class based on s_t-th sample is given by

$$\hat{T}_g = \sum_{i \in s_t} \beta_i^{s_t} x_i \tag{6.7.1}$$

where, $\beta_i^{s_t}$ is the weight associated with the i-th unit if it occurs in the s_t-th sample. Therefore,

$$E\left(\hat{T}_g\right) = \sum_{i=1}^{N} x_i \, \Sigma' \left\{ \beta_i^{s_t} \, p(s_t) \right\}$$

where, second summation Σ' is over all samples containing i-th unit. In order that $\hat{T}_g$ be an unbiased estimator of T, we have

$$\Sigma' \left(\beta_i^{s_t} \, p\,(s_t) \right) = 1 \tag{6.7.2}$$

for $i = 1, 2, \ldots, N$.

Let,

$$\beta_i^{s_t} = \frac{1}{\nu_i \, p(s_t)} \qquad \text{for all i}$$

where, $\nu_i (> 1)$ is the number of samples with non-zero probabilities, in which i-th unit occurs. Then this set of $\beta_i^{s_t}$ obviously satisfies the above N equations and so the class under consideration is non-empty.

For sampling without replacement, $\nu_i = n!\binom{N-1}{n-1}$ for all i. Thus, for this case

$$\hat{T}_g = \frac{\left(\sum_{r=1}^{n} X_r\right)}{n!\binom{N-1}{n-1}p(s_t)} \tag{6.7.3}$$

This is an ordered estimator in T_3-class which when unordered gives the same estimator as obtained by Tikkiwal (1966, Eq. (1.13) p. 67). This unordered estimator is

$$\frac{\sum_{r=1}^{n} X_r}{\binom{N-1}{n-1}P(S_p)} \tag{6.7.4}$$

where, $P(S_p)$ denotes the probability of the set S_p containing n! samples with distinct units as in the given s_t. For SRSWOR, the estimator reduces to the classical estimator $N(\sum X_r)/n$.

For sampling with replacement, given i-th unit in the sample can repeat at j places. Therefore,

$$\begin{aligned}\nu_i &= \sum_{j=1}^{n}\binom{n}{j}(N-1)^{n-j}\\ &= (N-1+1)^n - (N-1)^n\\ &= N^n - (N-1)^n.\end{aligned}$$

Thus,

$$\hat{T}_g = \frac{\sum_{i\in s_t} x_i}{\left[N^n - (N-1)^n\right]p(s_t)} \tag{6.7.5}$$

This estimator can further be improved upon by using the two techniques given in Chapter 1. For simple random sampling with replacement, the estimator reduces to

$$\frac{\sum_{i\in s_t} x_i}{\left[1-\left(1-\frac{1}{N}\right)^n\right]} \tag{6.7.6}$$

According to Godambe (1955, section 3, pp. 271-273) the said estimator is expected to be less efficient than the classical estimator ($\Sigma X_r/n$) for population with small coefficient of variation. However, before we decide to use one such estimator, we should also take into account another estimator $N\,\overline{x}_k$ given by Eq. (1.9.2) of Chapter 1.

We obtain below another useful estimator in the process of minimisation of variance of $\hat{T}_g$ subject to the conditions of its unbiasedness.

$$V\left(\hat{T}_g\right) = \sum_{s_t} \left\{ \sum_{i \in s_t} \beta_i^{s_t} x_i \right\}^2 p(s_t) - T^2 \tag{6.7.7}$$

Let

$$\phi = \sum_{s_t} \left\{ \sum_{i \in s_t} \beta_i^{s_t} x_i \right\}^2 p(s_t) \; - T^2 - 2 \sum_{i=1}^{N} \mu_i \left\{ \Sigma' \; \beta_i^{s_t} \; p(s_t) \right\}$$

By equating $\partial\phi/\partial\,\beta_i^{s_t}$ to zero, we get

$$\left\{ \sum_{i \in s_t} \beta_i^{s_t} x_i \right\} p(s_t)\, x_i = \mu_i \; p(s_t)$$

or,

$$\left(\hat{T}_g\right) p(s_t)\, x_i = \mu_i \; p(s_t) \tag{6.7.8}$$

Multiplying both the sides of the above equation by $\beta_i^{s_t}$ and then summing over all those samples containing i-th unit, we get

$$x_i \sum_{s_t \supset i} \left[\left(\hat{T}_g\right)_{s_t} p(s_t)\, \beta_i^{s_t} = \mu_i \right] \tag{6.7.9}$$

after using the conditions of unbiasedness for the i-th unit from Eq. (6.6.2). Let

$$k_i = \sum_{s_t \supset i} \left[\left(\hat{T}_g\right)_{s_t} p(s_t)\, \beta_i^{s_t} \right] \tag{6.7.10}$$

Then

$$x_i\, k_i = \mu_i$$

Therefore from Eq. (6.7.8)

$$\left(\hat{T}_g\right)_{s_t} = k_i$$

for all $s_t \supset i$. Consider any two such samples s_t and $s_{t'}$.

Then

$$\left(\hat{T}_g\right)_{s_t} = \left(\hat{T}_g\right)_{s_{t'}}$$

or,

$$x_i \left(\beta_i^{s_t} - \beta_i^{s_{t'}}\right) + \sum_{j(\neq i)} \beta_j^{s_t} x_j - \sum_{j'(\neq i)} \beta_{j'}^{s_{t'}} x_{j'} = 0$$

Since this equation is to hold good for any given population, let $x_i = 1$ and $x_j = 0$ for $j \neq i$. Then,

$$\beta_i^{s_t} = \beta_i^{s_{t'}}$$

Since this is true for any arbitrary s_t and $s_{t'}$ containing i-th unit, $\beta_i^{s_t} = C_i$. Substituting this in Eq. (6.7.2), we get

$$\beta_i^{s_t} = \frac{1}{P(i)} \tag{6.7.11}$$

where, P(i) is the probability of i-th unit being included in the sample. Thus,

$$\hat{T}_g = \sum_{i \in s_t} \frac{x_i}{P(i)} \tag{6.7.12}$$

For SWOR, it is the Horvitz-Thompson's estimator given in Chapter 3. However, for sampling with replacement, this estimator is different from that considered in Chapter 3. For some special populations and for specified probabilities p_i, Tikkiwal (1966) has shown that such estimators can be poorer than even the classical estimator for SWOR scheme. Therefore, there is need to investigate under what choice of p_i's, it can provide optimum results. For this, consider any two samples s_t and $s_{t'}$ containing i-th unit and some other (n–2) units, the only differing units being $i_j \in s_t$ and $i_{j'} \in s_{t'}$. Let $p(s_t)$ and $p(s_{t'})$ be not zero. Then,

$$\left(\hat{T}_g\right)_{s_t} = \left(\hat{T}_g\right)_{s_{t'}}$$

gives

$$\frac{x_{i_j}}{P(i_j)} = \frac{x_{i_{j'}}}{P(i_{j'})}$$

Considering such paired samples, we get

$$\frac{x_1}{P(1)} = \frac{x_2}{P(2)} = \ldots = \frac{x_N}{P(N)} = \frac{T}{\sum_{i=1}^{N} P(i)}$$

which gives

$$\frac{P(i)}{\sum_i P(i)} = \frac{x_i}{T} \tag{6.7.13}$$

for i = 1, 2 ,. . ., N. For sampling without replacement, this equation gives $P(i) = (n x_i)/T$, a result due to Horvitz and Thompson (1952, Eq. 7, p. 669). For these values of P(i)

$$\begin{aligned} \hat{T}_g &= \sum_{i \in s_t} \left[\frac{x_i}{\sum_i P(i)} \frac{T}{x_i} \right] \\ &= \frac{T}{\sum_i P(i)} \sum_{i \in s_t} (1) \qquad (6.7.14) \\ &= \frac{kT}{\sum P(i)} \end{aligned}$$

where, k denotes the number of distinct units in s_t.

For sampling without replacement, $\sum_i P(i) = n$ and k = n. Therefore, $\hat{T}_g = T$ and so $V(\hat{T}_g) = 0$. For sampling with replacement,

$$\sum_{i=1}^{N} P(i) \le \sum_{i=1}^{N} \sum_{r=1}^{n} (P_{ir}) = n$$

Therefore,

$$\sum_i P(i) \le n$$

and is a fixed quantity for a given sample design. However, k is a random variable. Therefore, $\hat{T}_g \neq T$ and so $V(\hat{T}_g) \neq 0$. Thus, we should prefer here sampling without replacement over sampling with replacement. This should be true even when the cost of enumerating the units is taken into account [Seth and Rao (1964)]. However, this requires a detail examination.

Since x_i's are not generally known, we should acquire advance knowledge of an auxiliary variable y to determine its functional relation with x, say x = f(y). Then choosing appropriate value of p_i, we may adopt varying probabilities scheme without replacement and then use the estimator due to Horvitz-Thompson. This should give near optimum estimation procedure, depending upon how reasonably good estimate of the function f(y) we make.

While discussing T_2-class estimation in Chapter 3, we have compared sampling without replacement schemes with sampling with replacement schemes. The above discussion throws further light on it.

EXERCISES

6.1. Define T_5-class of linear estimators for sampling with varying probabilities with or without replacement given by Bhargava and Tikkiwal (1978) and show that MVLUE does not exist in this class. Further discuss how the T_5-class given by the two authors differ than the one defined by Koop (1961).

6.2. Define T_6-class of linear estimators for sampling with varying probabilities with or without replacement given by Bhargava and Tikkiwal (1978), and show that under SRSWOR scheme MVLUE exists in this class. How this class is different than the one defined by Koop (1961)?

6.3. Examine the existence of MVLUE in T_6-class due to Bhargava and Tikkiwal for Midzuno systems of sampling without replacement for a special case taking N=3 and n=2.

6.4. Examine the existence MVLUE in T_6-class due to Bhargava and Tikkiwal for Horvitz-Thompson sampling scheme for a special case taking N=3 and n=2.

6.5. Define Godambe's general class of linear estimators for sampling with varying probabilities with or without replacement, and show that MVLUE does not exist in this class for any sampling design, except the uni-cluster design. Also show that Godambe's general class lies in $\hat{T}_7$-class.

[Godambe (1955), Lanke (1974, 1975)]

CHAPTER 7

Theory of Successive Sampling

7.1 INTRODUCTION

When a certain character or a group of characters of a population, changing over time, is studied on several occasions it is desirable to consider the use of ancillary information collected on the previous occasion to obtain improved estimates for the current occasion. This improvement is brought about in two different ways : (i) the ancillary information as obtained previously on a character is used to evolve a satisfactory scheme of sampling as in the survey of the acreage under jute in the State of Bengal (Mahalanobis, 1939); (ii) it is used to improve the procedure adopted to estimate the parametric values from the sample.

The theory of successive sampling has been developed mainly by Jessen (1942), Yates (1949), Patterson (1950) and Tikkiwal (1950, 53, 55, 56, 58, 67, 79) and Prabhu Ajgaonkar (1962). The most general situation considered so far is one that of estimation of the population mean of a time-conditioned finite population on successive occasions. Tikkiwal (1950, 55) has given, for such a situation, under a specified sampling scheme and a correlation pattern the MVLUE of the population mean on a particular occasion utilizing the information on this occasion and on all earlier occasions. His product correlation pattern is more general than the exponential one of Yates and Patterson (Tikkiwal, 1967 section 3, p. 248).

7.2 PRELIMINARY RESULTS

To have some insight in the theory, we present the main results for sampling on h successive occasions. We require here certain preliminary results, which are given in the form of Lemmas without proof (Tikkiwal, 1960). These results have been presented by the author (1986) in his lecture notes on a basic course on Theory of Sampling. They are included again for ready reference.

Lemma 7.2.1

Let $X_1, X_2, X_3, \ldots, X_p$ be p-variates with their first and second moments finite. Without any loss of generality, let

$E(X_i) = 0$ for $i = 1, 2, \ldots, p$. A linear bond among $X_1, X_2, \ldots, X_p$ is said to exist if $\sum C_i X_i = 0, \sum C_i^2 \neq 0$. A set of r linear bonds among $X_1, X_2, \ldots, X_p$ is said to be complete, if all the r bonds of the set are independent and any other linear bond among $X_1, X_2, \ldots, X_p$ depends linearly on these r bonds. The rank, R(F) of the joint distribution F $(x_1, x_2, \ldots, x_p)$ of $X_1, X_2, \ldots, X_p$ is then (p-r) by definition. When R(F) = p, the distribution is said to be non-degenerate.

Lemma 7.2.2

The rank R(F) of distribution F $(x_1, x_2, \ldots, x_p)$ is equal to the rank of the covariance-matrix $\|\mu_{ij}\|$, where $\mu_{ij} = E(X_i X_j)$ for i, j = 1, 2, . . ., p.

Corollary 7.2.2

The distribution of F is non-degenerate iff the covariance matrix $\| \mu_{ij} \|$ is positive definite.

Lemma 7.2.3

Let there be a non-degenerate p-variate population with first and second moments finite. Let X_{ir} denote the variate value of the unit drawn at the r-th draw for i-th variate for r = 1, 2, . . ., n_i and i = 1, 2,. . ., p. Let $e_\alpha = \sum_{i=1}^{p} \sum_{r=1}^{n_i} w_{ir} X_{ir}$, with w_{ir} as weights, be a linear unbiased estimator of population mean μ_α for the α-th variate based on all $\sum_{i=1}^{p} n_i$ observations, then e_α is the MVLUE of μ_α among the class of linear unbiased estimators if and only if

$$Cov(X_{ir}, e_\alpha) = K_{i\alpha}, \quad r = 1, 2, \ldots, n_i; \quad i=1,2, \ldots, p$$

Corollary 7.2.3

If the weights w_{ir} are not estimated from the sample and so non-stochastic in nature, then

$$Var(e_\alpha) = K_{\alpha\alpha} = Cov(X_{\alpha r}, e_\alpha)$$

Lemma 7.2.4

Let $\bar{x}$ and $\bar{y}$ be the sample means for the variates X and Y based on observations for the two variates on the same n units selected at random without replacement from the population of size N, then

$$\text{Cov}(\bar{x}, \bar{y}) = \text{Cov}(X_r, \bar{y}) = \text{Cov}(Y_r, \bar{x})$$

$$= \frac{N-n}{Nn} \rho S_x S_y$$

where X_r and Y_r denote the outcomes at the r-th draw respectively for the variates X and Y and

$$S_z^2 = \sum_{i=1}^{N} (z_i - \mu_z)^2 / (N-1) \text{ with } \mu_z \text{ being equal to}$$

$$\left(\sum_{i=1}^{N} z_i\right)/N \quad \text{for } z = x, y \text{ and}$$

$$\rho = \sum_{i=1}^{N} (x_i - \mu_x)(y_i - \mu_y) / \{(N-1) S_x S_y\}$$

If the sample means for the variate X is based on n_x units and the sample mean for Y on n_y different units, then each of the covariance terms is given by the expression $-\rho S_x S_y / N$. This expression also determines $\text{Cov}(X_r, Y_s)$ for $r \neq s$.

7.3 SUCCESSIVE SAMPLING ON h (≥2) OCCASIONS

Let a population π be observed for the variate X, as follows, on occasions 1 to h to estimate the population mean μ_h on the h-th occasion. On the first occasion, n_1 units are drawn out of N with SRSWOR. On the second occasion n_2' units from n_1 units with SRSWOR are retained from n_1 and n_2'' units selected with SRSWOR fron $N-n_1$ units which are not observed on the first occasion. Proceeding in this manner, on t-th occasion (for $t \geq 2$) n_t' units are retained from n_{t-1} units taken on (t–1)-th occasion and supplemented by n_t'' new units with SRSWOR scheme from $N - n_1 - n_2'' - n_3'' - \ldots - n_{t-1}'' = N - n_1 - \sum_{i=2}^{t-1} n_i''$ units.

Let us observe the variate value for x on each unit drawn on an occasion. Let X_{ir} denote the value of the unit drawn at the r-th draw on i-th occasion, for all possible r and i. The correlation between generated random variates at a particular draw on the i-th and j-th occasions is assumed to be $\rho_{ij} = \underset{t=1}{\overset{j-1}{\pi}} \rho_{t,t+1}$ where $\rho_{t,t+1}$ is the correlation between the units on t-th and (t + 1)-th occasions. Then the MVLUE $\hat{X}_h$ (Tikkiwal, 1950, 67) of the population mean μ_h for h $\geq$ 2 is

$$\hat{X}_h = (1-\phi_h) Z_h + \phi_h X_{hn_h''} \qquad (7.3.1)$$

$$Z_h = X_{hn_h'} + \beta_{h-1,h} (\hat{X}_{h-1} - X_{h-1n_h'})$$

where,

$X_{hn_h'}$ = sample mean, on the h-th occasion of n_h' units common between (h–1)-th and h-th occasions

$X_{h-1n_h'}$ = sample mean, on the (h–1)-th occasion, of the same n_h' units

$X_{hn_h''}$ = sample mean, on the h-th occasion, of the n_h'' units observed for the first time on the h-th occasion

$\beta_{h-1,h} = \rho_{h-1,h} \dfrac{S_h}{S_{h-1}}$ = Regression coefficient of X_h on X_{h-1}

The weighting coefficient ϕ_h is given by

$$\frac{\phi_h}{1-\phi_h} = \frac{n_h''}{n_h'}(1-\rho^2_{h-1,h}) + \rho^2_{h-1,h} \frac{n_h''}{n_{h-1}''} \phi_{h-1}$$

and $\phi_1 = \dfrac{n_1''}{n_1}$ where $n_1'' = n_1 - n_2'$.

Also $S_i^2 = \dfrac{1}{N-1} \sum\limits_{j=1}^{N} (x_{ij} - \mu_i)^2$ for i = 1, 2,..., h.

Thus we can write (7.3.1) as

$$\hat{X}_h = (1-\phi_h)\left[X_{hn_h'} + \rho_{h-1,h} \frac{S_h}{S_{h-1}} (\hat{X}_{h-1} - X_{h-1n_h'})\right] + \phi_h X_{hn_h''} \qquad (7.3.2)$$

For h=1, the sample mean $\overline{X}$, based on n_1 units is easily seen to be MVLUE. We prove the following theorem.

Theorem 7.3.1

$\hat{X}_h$ is MVLUE for μ_h and its variance is given by

$$\text{Var} \quad (\hat{X}_h) = S_h^2 \left[\frac{\phi_h}{n_h''} - \frac{1}{N} \right]$$

Proof :

From Eq. (7.3.2), it is clear that $\hat{X}_h$ is a linear estimator provided $\hat{X}_{h-1}$ is so. In turn $\hat{X}_{h-1}$ is linear provided that $\hat{X}_{h-2}$ is so. Proceeding in this manner, we see that $\hat{X}_2$ is a linear estimator provided $\hat{X}_1$ is so. This is true, therefore, $\hat{X}_h$ is a linear estimator.

In order to examine the unbiasedness of $\hat{X}_h$, we note

$$E(\hat{X}_h) = (1 - \phi_h)\, E(Z_h) + \phi_h\, E(X_{hn_h''})$$

$$= (1 - \phi_h) \left[E(X_{hn_h'}) + \rho_{h-1,h} \frac{S_h}{S_{h-1}} E(\hat{X}_{h-1} - X_{h-1n_h'}) \right] + \phi_h\, E(X_{hn_h''})$$

$$= (1 - \phi_h) \left[\mu_h + \rho_{h-1,h} \frac{S_h}{S_{h-1}} (\mu_{h-1} - \mu_{h-1}) \right] + \phi_h\, \mu_h$$

$$= \mu_h \qquad (7.3.3)$$

provided $E(\hat{X}_{h-1}) = \mu_{h-1}$. Similarly from Eq. (7.3.2) we observe that $E(\hat{X}_{h-1}) = \mu_{h-1}$, provided $E(\hat{X}_{h-2}) = \mu_{h-2}$. Proceeding in this way, we get $E(\hat{X}_h) = \mu_h$, provided ultimately $E(\hat{X}_1) = \mu_1$, which is true. Hence $\hat{X}_h$ is unbiased linear estimator for the population mean on the h-th occasion.

In order to prove that $\hat{X}_h$ is MVLUE, we now make use of Lemma 7.2.3 as follows.

Case I

Let X_{hr} be the variate value of the unit drawn at r-th draw on h-th occasion for $r = 1,2, \ldots, n_h'$. Then,

$$\text{Cov}(X_{hr}, \hat{X}_h) = (1-\phi_h)\left[\text{Cov}(X_{hr}, X_{hn_h'}) + \rho_{h-1,h}\frac{S_h}{S_{h-1}}\left\{\text{Cov}(X_{hr}, \hat{X}_{h-1}) - \text{Cov}(X_{hr}, X_{h-1n_h'})\right\}\right] + \phi_h\,\text{Cov}(X_{hr}, X_{hn_h''}) \qquad (7.3.4)$$

By Lemma 7.2.4, we have

$$\text{Cov}(X_{hr}, X_{hn_h'}) = \frac{N-n_h'}{Nn_h'} S_h^2 \qquad (7.3.5)$$

$$\text{Cov}(X_{hr}, X_{h-1n_h'}) = \rho_{h-1,h} S_h S_{h-1} \frac{N-n_h'}{Nn_h'} \qquad (7.3.6)$$

and $$\text{Cov}(X_{hr}, X_{hn_h''}) = -\frac{S_h^2}{N} \qquad (7.3.7)$$

Substituting the values from Eqs. (7.3.5), (7.3.6) and (7.3.7) in Eq. (7.3.4), we get

$$\text{Cov}(X_h, \hat{X}_h) = (1-\phi_h)\left[\frac{N-n_h'}{Nn_h'}(1-\rho^2_{h-1,h}) S_h^2\right] - \frac{1}{N}\phi_h S_h^2 + (1-\phi_h)\,\rho_{h-1,h}\frac{S_h}{S_{h-1}}\text{Cov}(X_{hr}, \hat{X}_{h-1}) \qquad (7.3.8)$$

Since $$\hat{X}_{h-1} = \sum_{i=1}^{h-1}\sum_{r=1}^{n_i} w_{ir} X_{ir}$$

$$\text{Cov}(X_{hr}, \hat{X}_{h-1}) = \sum_{i=1}^{h-1}\sum_{r=1}^{n_i} w_{ir}\,\text{Cov}(X_{ir}, X_{hr})$$

$$= \sum_{i=1}^{h-1}\sum_{r=1}^{n_i} w_{ir}\,\rho_{h-1,h}\frac{S_h}{S_{h-1}}\text{Cov}(X_{ir}, X_{h-1r})$$

$$= \rho_{h-1,h}\frac{S_h}{S_{h-1}}\sum_{i=1}^{h-1}\sum_{r=1}^{n_i} w_{ir}\,\text{Cov}(X_{ir}, X_{h-1r})$$

$$= \rho_{h-1,h} \frac{S_h}{S_{h-1}} \text{Cov}(X_{h-1r}, \hat{X}_{h-1}) \tag{7.3.9}$$

Assuming Theorem 7.3.1 to be true for h-1 (see also, Fisher, 1925) we get

$$\text{Cov}(X_{h-1r}, \hat{X}_{h-1}) = \text{Var}(\hat{X}_{h-1})$$

$$= S_{h-1}^2 \left[\frac{\phi_{h-1}}{n''_{h-1}} - \frac{1}{N}\right] \tag{7.3.10}$$

Thus from Eqs. (7.3.8) and (7.3.10) we get,

$$\text{Cov}(X_{hr}, \hat{X}_{h-1}) = (1-\phi_h)\left[\frac{N - n'_h}{Nn'_h}(1-\rho_{h-1,h}^2) S_h^2\right]$$

$$- \phi_h \frac{S_h^2}{N} + (1-\phi_h)\, \rho_{h-1,h}^2\, S_h^2 \left[\frac{\phi_{h-1}}{n''_{h-1}} - \frac{1}{N}\right]$$

$$= S_h^2\,(1-\phi_h)\left[\frac{1-\rho_{h-1,h}^2}{n'_h} + \rho_{h-1,h}^2\left(\frac{\phi_{h-1}}{n''_{h-1}}\right)\right] - \frac{S_h^2}{N}$$

Using definition of ϕ_h,

$$= S_h^2\left[\frac{\phi_h}{n''_h} - \frac{1}{N}\right] \tag{7.3.11}$$

Case II

Let X_{hr} be the variate value of the unit drawn at the r-th draw on h-th occasion for r = 1, 2, . . ., n''_h. Then

$$\text{Cov}(X_{hr}, \hat{X}_{h-1}) = (1-\phi_h)\left[\text{Cov}(X_{hr}, X_{hn'_h}) + \rho_{h-1,h}\frac{S_h}{S_{h-1}}\right.$$

$$\left.\left\{\text{Cov}(X_{hr}, \hat{X}_{h-1}) - \text{Cov}(X_{hr}, X_{h-1n'_h})\right\}\right] + \phi_h\, \text{Cov}(X_{hr}, X_{hn''_h}) \tag{7.3.12}$$

From Lemma 7.2.4, we get

$$\text{Cov}(X_{hr}, X_{hn'_h}) = -\frac{S_h^2}{N} \tag{7.3.13}$$

and,

$$\text{Cov}\,(X_{hr}, X_{hn_h''}) = \frac{N - n_h''}{N n_h''} S_h^2 \tag{7.3.14}$$

and,

$$\text{Cov}\,(X_{hr}, X_{h\text{-}1n_h'}) = -\frac{\rho_{h\text{-}1,h}\, S_h\, S_{h\text{-}1}}{N} \tag{7.3.15}$$

Putting the values from Eqs. (7.3.13), (7.3.14), (7.3.15) in (7.3.12) we get

$$\text{Cov}\,(X_{hr}, \hat{X}_h) = (1-\phi_h)\left[\frac{1}{n}(\rho_{h\text{-}1,h}^2 - 1)\, S_h^2 + \rho_{h\text{-}1,h}\frac{S_h}{S_{h\text{-}1}}\text{Cov}\,(X_{hr}, \hat{X}_{h\text{-}1})\right]$$

$$+\phi_h \frac{N - n_h''}{N n_h''} S_h^2 \tag{7.3.16}$$

Now we find $\text{Cov}\,(X_{hr}, \hat{X}_{h\text{-}1})$

As before, we note that

$$\text{Cov}\,(X_{hr}, \hat{X}_{h-1}) = \sum_i \sum_r w_{ir}\, \text{Cov}\,(X_{hr}, X_{ir})$$

$$= \sum_i \sum_r w_{ir}\left(-\rho_{i,h}\frac{S_i S_h}{N}\right)$$

Noting that $\hat{X}_{h-1}$ is a linear unbiased estimator

$$= -\rho_{h-1,h}\frac{S_{h-1} S_h}{N} \tag{7.3.17}$$

Therefore,

$$\text{Cov}\,(X_{hr}, \hat{X}_h) = (1-\phi_h)\left[-\frac{S_h^2}{N} + \rho_{h-1,h}^2\frac{S_h^2}{N} - \rho_{h-1,h}^2\frac{S_h^2}{N}\right] + \phi_h \frac{N - n_h''}{N n_h''} S_h^2$$

$$= S_h^2\left[\frac{\phi_h}{n_h''} - \frac{1}{N}\right]$$

same as in case I.

We now proceed to find $\text{Cov}\,(X_{ir}, \hat{X}_h)$ for $i = 1, 2, \ldots, h-1$ where X_{ir} denote the variate value of the unit drawn at the r-th draw on i-th occasion for r = 1, 2, . . ., n_i, where $n_i = n_i' + n_i''$.

Case I

When the unit drawn at r-th draw on i-th occasion is retained on h-th occasion, then

$$\mathrm{Cov}\,(X_{ir}, \hat{X}_h) = (1-\phi_h)\left[\mathrm{Cov}\,(X_{ir}, X_{hn'_h}) + \rho_{h-1,h}\frac{S_h}{S_{h-1}}\left\{\mathrm{Cov}\,(X_{ir}, \hat{X}_{h-1}) - \mathrm{Cov}\,(X_{ir}, X_{h-1n'_h})\right\}\right] + \phi_h\,\mathrm{Cov}\,(X_{ir}.\,X_{hn''_h}) \tag{7.3.18}$$

Using Lemma 7.2.4, we get

$$\mathrm{Cov}\,(X_{ir}, \hat{X}_h) = (1-\phi_h)\left[\rho_{i,h} S_i S_h \frac{N-n'_h}{Nn'_h} + \rho_{h-1,h}\frac{S_h}{S_{h-1}}\left\{\mathrm{Cov}\,(X_{ir}, \hat{X}_{h-1}) - \rho_{i,h-1}\frac{N-n'_h}{Nn'_h} S_i S_{h-1}\right\}\right] - \phi_h\,\rho_{i,h}\frac{S_i S_h}{N}$$

Since $\rho_{i,h} = \rho_{h-1,h}\,\rho_{i,h-1}$, therefore

$$\mathrm{Cov}\,(X_{ir}, \hat{X}_h) = (1-\phi_h)\left[\rho_{i,h}\, S_i S_h \frac{N-n'_h}{Nn'_h} + \rho_{h-1,h}\frac{S_h}{S_{h-1}}\mathrm{Cov}\,(X_{ir}, \hat{X}_{h-1}) - \rho_{i,h} S_i S_h \frac{N-n'_h}{Nn'_h}\right] - \phi_h\,\rho_{i,h}\frac{S_i S_h}{N}$$

$$= (1-\phi_h)\left[\rho_{h-1,h}\frac{S_h}{S_{h-1}}\mathrm{Cov}\,(X_{ir}, \hat{X}_{h-1})\right] - \phi_h\,\rho_{i,h}\frac{S_i S_h}{N} \tag{7.3.19}$$

Case II

When the unit drawn on i-th occasion at r-th draw is not retained on h-th occasion, then proceeding as before in case I, we note that $\mathrm{Cov}\,(X_{ir}, \hat{X}_h)$ is the same as given by Eq. (7.3.19). After substituting the value of $\mathrm{Cov}\,(X_{ir}, \hat{X}_{h-1})$ from this very equation, we get

$$\text{Cov}(X_{ir}, \hat{X}_h) = (1-\phi_h)\,\rho_{h-1,h}\,\frac{S_h}{S_{h-1}}\left[(1-\phi_{h-1})\,\rho_{h-2,h-1}\,\frac{S_{h-1}}{S_{h-2}}\right.$$

$$\left.\text{Cov}(X_{ir}, \hat{X}_{h-2}) - \phi_{h-1}\,\rho_{i,h-1}\,\frac{S_i\,S_{h-1}}{N}\right] - \phi_h\,\rho_{i,h}\,\frac{S_i\,S_h}{N}$$

$$= (1-\phi_h)(1-\phi_{h-1})\,\rho_{h,h-2}\,\frac{S_h}{S_{h-2}}\,\text{Cov}(X_{ir}, \hat{X}_{h-2})$$

$$-(1-\phi_h)\,\phi_{h-1}\,\rho_{i,h}\,\frac{S_i S_h}{N} - \phi_h\,\rho_{i,h}\,\frac{S_i S_h}{N} \qquad (7.3.20)$$

Again substituting the value of $\text{Cov}(X_{ir}, \hat{X}_{h-2})$ in Eq. (7.3.20) from Eq. (7.3.19)

$$= (1-\phi_h)(1-\phi_{h-1})(1-\phi_{h-2})\,\rho_{h,h-3}\,\frac{S_h}{S_{h-3}}\,\text{Cov}(X_{ir}, \hat{X}_{h-3})$$

$$-(1-\phi_h)(1-\phi_{h-1})\,\phi_{h-2}\,\rho_{i,h}\,\frac{S_i S_h}{N} - (1-\phi_h)\,\phi_{h-1}\,\rho_{i,h}\,\frac{S_i S_h}{N}$$

$$-\phi_h\,\rho_{i,h}\,\frac{S_i S_h}{N}.$$

Proceeding further in this way

$$= \mathop{\pi}_{t=i+1}^{h}(1-\phi_t)\,\rho_{i,h}\,\frac{S_h}{S_i}\,\text{Cov}(X_{ir}, \hat{X}_i) - \{(1-\phi_h)(1-\phi_{h-1})\ldots.$$

$$(1-\phi_{i+2})\,\phi_{i+1} + \ldots + (1-\phi_h)(\phi_{h-1}) + \phi_h\}\,\rho_{i,h}\,\frac{S_i S_h}{N} \qquad (7.3.21)$$

Assuming $\hat{X}_i$ to be MVLUE for i-th occasion we can write Eq. (7.3.21) as

$$\text{Cov}(X_{ir}, \hat{X}_h) = \mathop{\pi}_{t=i+1}^{h}(1-\phi_t)\,\rho_{i,h}\,\frac{S_h}{S_i}\,S_i^2\left[\frac{\phi_i}{n_i''} - \frac{1}{N}\right]$$

$$-\{(1-\phi_h)(1-\phi_{h-1})\ldots(1-\phi_{i+2})\phi_{i+1} + \ldots + (1-\phi_h)\phi_{h-1} + \phi_h\}\,\rho_{i,h}\,\frac{S_i S_h}{N} \qquad (7.3.22)$$

Clearly, we note that R.H.S. of $\text{Cov}(X_{ir}, \hat{X}_h)$ in Eq. (7.3.22) is independent of r. Thus $\hat{X}_h$ satisfies the condition of MVLUE given in Lemma 7.2.3 provided that $\hat{X}_{h-1}, \hat{X}_{h-2}, \ldots, \hat{X}_2$ are MVLUE's. Thus if we know that $\hat{X}_2$ is MVLUE then it follows that $\hat{X}_h$ is MVLUE of μ_h for all $h \geq 1$. We note that

$$\hat{X}_2 = (1-\phi_2)\left[X_{2n_2'} + \rho_{1,2}\frac{S_2}{S_1}(\hat{X}_1 - X_{1n_2'})\right] + \phi_2 X_{2n_2''}$$

where,

$$\frac{\phi_2}{1-\phi_2} = \frac{n_2''}{n_2'}(1-\rho_{1,2}^2) + \frac{n_2''}{n_1''}\rho_{1,2}^2\,\phi_1$$

Thus,

$$\text{Cov}(X_{ir}, \hat{X}_2) = \rho_{1,2} S_1 S_2\left[(1-\phi_2)\frac{\phi_1}{n_1''} - \frac{1}{N}\right]$$

and,

$$\text{Cov}(X_{2r}, \hat{X}_2) = S_2^2\left[\frac{\phi_2}{n_2''} - \frac{1}{N}\right] \qquad \text{for all r.}$$

Now by Lemma 7.2.3, $\hat{X}_2$ is MVLUE and therefore $\hat{X}_h$.

In practice, the various correlation and regression coefficients are not generally known in advance for any one of the h occasions after the first, occurring in estimator $\hat{X}_h$. Then the common practice is to estimate these coefficients from the common units in the sample on that occasion. Let $b_{h-1,h}$ and $\hat{\phi}_h$ be the estimators of $\beta_{h-1,h}$ and ϕ_h. Let the estimator of μ_h be

$$\hat{X}_h = (1-\hat{\phi}_h) Z_h' + \hat{\phi}_h X_{hn_h''} \tag{7.3.23}$$

where,

$$Z_h' = X_{hn_h'} + b_{h-1,h}(\hat{X}_{h-1} - X_{h-1n_h'})$$

When n_h' the common units on the h-th occasion, forms a sub-sample of n_{h-1}'' the new units on (h-1)-th occasion; the author (1956) has shown under normality assumption that the above estimator $\hat{X}_h$ is an unbiased estimator of μ_h and its variance is given by

$$\text{Var}(\hat{X}_h) = \frac{S_h^2}{n_h''} E(\hat{\phi}_h) \tag{7.3.24}$$

However, if the common units n'_h do not satisfy the above conditions, he has further shown that $\hat{X}_h$ is a consistent and asymptotically unbiased estimator of μ_h and that the mean square deviation of $\hat{X}_h$ about μ_h satisfies the inequality

$$\frac{\phi_h}{n''_h} S_h^2 \leq \text{Var}(\hat{X}_h) \leq \frac{E(\hat{\phi}_h) S_h^2}{n''_h} \tag{7.3.25}$$

The author also gives a consistent and asymptotically unbiased estimator of the variance or the mean square deviation of $\hat{X}_h$ as the case may be. This estimator is

$$\hat{V} = \hat{\phi}_h \frac{\hat{\sigma}_h^2}{n''_h} \tag{7.3.26}$$

where, $\hat{\sigma}_h^2$ is the usual estimator of σ_h^2 based on n_h units. Tikkiwal, G.C. and Gupta (1985) has examined the nature of approximation involved in the estimator of variance and come to the conclusion that the above estimator fairs well for large-size populations and large n. When it is not so, they still recommend the use of $\hat{V}$ under certain situations, where as for some other situations they suggest a modified estimator.

Sukhatmes, P.V., B.V. and S. and Ashok C. (1984) and Cochran (1977) give a fairly good treatment to the theory of successive sampling on h occasions. Konijn (1973, section 6, p. 411) gives the treatment to the case of sampling on two occasions, when population size changes from one occasion to the other. A reader is recommended to go through this material alongwith the material given here. These authors, however, have missed an important aspect of the theory, viz., the estimation of the variance of the estimator on an occasion; which is discussed above.

For the estimation of variance in case of finite population, one may go through the review paper by the author (1979, sections 2 and 3). In this paper, the author also discusses the validity of results in case of finite and infinite populations when the correlation pattern breaks down. Briefly it can be said that the expression for estimator of variance in (7.3.26) under normality assumption and the expression $\hat{\sigma}_h^2 [\hat{\phi}_h / n''_h - 1/N]$ as the estimator of variance in case of finite population are valid expressions to be used under variety of situations, when the estimator of population mean used is one given in expression (7.3.1).

7.4 A GENERAL DISCUSSION ON UNISTAGE SUCCESSIVE SAMPLING WITH VARYING PROBABILITIES

So far we have discussed the results when SRSWOR scheme is used in selecting a sample from the finite or infinite populations on all the occasions. In some sample surveys, however, there are situations where sample units should be selected with varying probabilities. The work done to meet such situations is reviewed below.

Des Raj (1965) considered the following sampling scheme over two occasions. On the first occasion a sample s_1 of n unit is selected with probabilities proportional to sizes x_i attached to the units and with replacement. On the second occasion a simple random sub-sample s_1' of $m = n\lambda$ units from s_1 without replacement and an independent pps sample s_2' of $n\mu = (n - m)$ units with replacement from the whole population are selected. He obtained an unbiased estimator of the population total on second occasion considering a difference estimator of the population total on the matched part. However, Pathak and Rao (1967) proved the inadmissibility of Des Raj's estimator. Certain modifications to Pathak and Rao's estimator for the population total on the second occasion, were suggested by Singh, R. (1972). Tripathi (1970) extended the results due to Des Raj to $h (\geq 2)$ occasions.

Ghangurde and Rao (1969) improved up on the results of Des Raj by adopting two separate techniques of selecting units : (i) Steven's IPPS (Inclusion Probability Proportional to Size) and (ii) Rao-Hartley-Cochran technique on the first occasion. The authors showed that the estimator of the population mean in each case is more efficient than Des Raj's estimator. Jayanti Chotai (1974) further provided some improvements by adopting pps sampling scheme without replacement on two occasions. Avadhani and Srivastava (1972) developed the results for sampling on two occasions like Des Raj but by adopting Midzuno-Sen and Lahiri's method of selection on the first occasion instead of pps with replacement. However, the efficiencies of their methods were not compared with those of the methods due to Des Raj, Ghangurde and Rao.

Avadhani and Sukhatme (1970) used the information of first occasion in two different ways : (i) While selecting a sub-sample of units on the second occasion, and (ii) At estimation stage. A sub-sample of units, from the units selected on the first occasion, is selected by the Rao-Hartley and Cochran (1962) technique using the variate values of the units observed on the first occasion in forming probabilities required in the technique. Their estimator based on the variate values, of the matched units, on the second occasion is then used. To use the information at estimation stage, a random sub-sample, from the units selected on the first occasion is selected by SRSWOR technique and a ratio-estimator is built-up based on matched units and the units on the first occasion. Avadhani and Sukhatme then combine these two estimators separately with the simple mean based on new units using minimum variance combination. Thus, they obtain two estimates of the population mean on the second occasion. Their relative efficiencies are compared under the following model :

$$y_i = \beta x_i + e_i \ , i = 1, 2, \ldots, N$$

where,

$$\sum_{i=1}^{N} e_i = \sum_{i=1}^{N} e_i x_i = 0, \text{ and } e_i^2 = a x_i^g$$

with $a > 0$ and $0 \leq g \leq 2$. It is shown that the estimator on the second occasion using ratio estimator for the matched part is, more or less, equally efficient than the estimator using Rao, Hartley and Cochran's estimator according as g is less or greater than or equal to unity. Avadhani and Sukhatme (1972 a) further show that their estimator is more efficient, under the above model, than that of Pathak and Rao (1967). The two authors (1972 b), further, extended the theory of fps sampling for more than two occasions.

7.5 A GENERAL DISCUSSION ON MULTISTAGE SUCCESSIVE SAMPLING

The work discussed in earlier Sections is on unit-stage successive sampling. In practice, however, the population generally consists of multi-stage units for which partial replacement of units from the sample drawn on the previous occasion can be done in a number of ways. The theory of two-stage successive sampling was first developed by Tikkiwal (1958). Tikkiwal (1964, 65) gave various lemmas useful for developing, in generality, the theory of two-stage sampling on successive occassions with equal or with varying probabilities and dealt with, for illustration, the theory, under equal probabilities system, when the first stage units (fsu's) alone are partially replaced on various occassions after the first by observing the same set of second-stage units (ssu's). A minimum variance linear unbiased estimator of the population mean on the h-th occasion was obtained under the following assumption :

$$\rho_{fii'} = \underset{t=i}{\overset{i'-1}{\pi}} \rho_{ft,t+1}$$

for all i and i′

$$\rho_{sii'} = \underset{t=1}{\overset{i'-1}{\pi}} \rho_{st,\ t+1}$$

$$\beta^f_{p,q} = \beta^s_{p,q} = \beta_{p,q}\ ;\ p, q = t-1, t \text{ and } t = 2,3,\ldots,h$$

where, $\rho_{fii'}$ and $\rho_{sii'}$ denote the correlation coefficients between the i-th and i'-th occasions at first and second stages and $\beta^f_{p,q}$ and $\beta^s_{p,q}$ denote the regression coefficients of the q-th occasion on the pth occasion at first and second stages respectively. Furhter it is shown that these assumptions are necessary and sufficient to obtain the MVLUE in the given form.

Singh (1968) obtained an estimator of the population mean on the second occasion under certain restrictive assumptions. This is a special case of Tikkiwal (1965) for h=2. Singh further obtained the estimators of the mean on the third occasion based on two sampling patterns of replacement of fsu's. Under sampling pattern I, a fraction p of n fsu's observed on the first occasion is retained on the second and third occasions and a fresh sample of nq fsu's, where p+q = 1, is randomly selected, on each of the two occasions, from the fsu's not selected upto that time. In sampling pattern II, the sampling scheme remains same on the first two occasions, but on third, the repeated units on the second occasion are not repeated again on this occasion but a fresh sub-sample of size np fsu's out of the nq (taking q > p) fsu's, selected afresh on the second occasion, is selected. It is shown that for estimating the mean on the third occasion, sampling pattern I should be used in preference to sampling pattern II. However for the estimation of mean over all the three occasions, sampling pattern II should be preferred. It is to be noted that both these sampling patterns satisfy Tikkiwal's condition (1965, section 3, p. 128).

A condition similar to Tikkiwal's condition for a sampling pattern in a multi-stage design to satisfy has been given by him earlier (1950, 1956) for uni-stage designs. That such conditions seem to be optimum follows from the discussions by Tikkiwal (1976, section 1).

Singh and Kathuria (1969) suggested four patterns of partial replacement of units in a two stage sampling design on h (≥ 2) occasions :

I. Use the same fsu's on all h occasions but from each fsu take a fresh sample of ssu's within them on each occasion after the first.

II. Use the same fsu's on h different occasions as in I but from each fsu, on all subsequent occasions, retain only a fraction p of the ssu's, taken on the first occasion, within them and select a fraction q of ssu's afresh on each of the (h–1) occasions.

III. Retain, on all subsequent occasions, only a fraction p of the fsu's, selected on the first occasion, with their samples of ssu's and select a fresh fraction q of fsu's (q + p = 1) on each of the (h–1) subsequent occasions.

IV. Retain a fraction p of fsu's as in III, but from each such fsu retain, on all subsequent occasions, only a fraction r of ssu's and select, on each of the (h–1) subsequent occasions, a fraction s of ssu's afresh (r + s = 1).

It may be noted that Singh's pattern I (1968) is a special case of Pattern III above for h = 3.

Estimators of the population mean and their variances were obtained by them on second occasion under the patterns of replacement II and III. A cost function suitable to the design for h = 2 was also suggested and applied to certain data. They also generalised these results for these two patterns, to $h(\geq 2)$ occasions under the assumptions that the pattern of variability between fsu's and ssu's remains the same from occasion to occasion and N and M are large.

Abraham, Khosla and Kathuria (1969) gave the two estimators and their variances on second occasion under above assumptions and replacement pattern IV. The efficiencies of the Estimators I and II over that of the sample mean per ssu on the second occasion were calculated for different values of the parameters involved in the estimators. With the help of these calculations, the authors concluded that the Estimator II is equal or more efficient than Estimator I for all different values of the parameters involved. An estimator of the change and an overall estimator of the mean over two occasions with their variances were also obtained. They used these estimators in incidence of pest and disease surveys for estimating (i) the mean incidence of pest and disease on the second occasion, (ii) the change in incidence of pest and disease from one year to another year and (iii) the overall estimator of mean incidence over two years. Srivastava and Singh (1974) showed that Table 2 [Abraham *et al.* (1969)] was wrong and gave the correct table showing to the contrary that

Estimator I is never less efficient than Estimator II. In fact, that Estimator II is not MVLUE is shown by the latter authors by using Tikkiwal's Lemmas (1964).

Kathuria and Singh (1971 a) compared the relative efficiencies of the two estimators of population mean for above sampling patterns II and III referred to above, upto five occasions. As a result of these comparisons, it was shown that none of the estimators is uniformly better than the other. Kathuria and Singh (1971 b) further compared the relative efficiencies of Estimator I [Abraham, Khosla and Kathuria (1969)], based on pattern IV, with respect to the two estimators based on patterns II and III on second occasion for different values of the parameters involved.

A conclusion, that can be drawn from these studies, is that for each specific situation some preliminary studies are necessary in order to find which pattern is to be adopted for that situation to attain optimality. Further, for variety of situations, patterns II and IV would have nearly the same efficiency.

The two authors further extended the Abraham, Khosla and Kathuria's results from two occasions to three occasions using sampling pattern IV as it is and two new sampling patterns, not covered in earlier studies, where, in each, partial retention is made among fsu's as well as ssu's. Under restrictive assumptions, one of the new patterns was shown to be superior than the other two. This study too confirms the above conclusion.

The above discussion confines to sampling upto 3 occasions. However, general results [Srivastava and Shivtar Singh (1974)] for $h \geq 2$ are available under specific replacement pattern, where partial retention is made among both fsu's and ssu's.

We now present results on two-stage successive sampling where auxiliary information is used either at the selection stage or at the estimation stage or at both. Tikkiwal's Lemma 2.2 (1964) can be useful for establishing optimal properties of the estimators in studies where auxiliary information is used at the selection stage.

Utilising the auxiliary information Shivtar Singh and Srivastava (1973) obtained the minimum variance linear unbiased estimators of (i) the population mean on the current occasion, (ii) the change in the population means from one occasion to the other and (iii) the population mean over all h occasions, for a two stage design in which fsu's are partially replaced and the ssu's in the retained fsu have been kept fixed. The entire study was though made under general correlation model, the expressions for various results are complicated.

Purakam and Koop (1966) studied to a considerable depth where fsu's are drawn with unequal probabilities with replacement and ssu's within them are selected with equal probabilities without replacement. To estimate the totals of the population characteristic under study and the changes in such totals, four types of estimators : (i) the linear composite estimator, (ii) the modified linear composite estimator, (iii) the general linear estimator and (iv) the ratio type composite estimator, are considered. The expressions for the variances of these estimatrs are given. The percentage gains in efficiency using such estimators over the simple unbiased estimator are also given. The estimation theory is discussed in detail for two-stage sampling design on two and three occasions and the extension to more than two stages as well as occasions is also indicated. The study is confined to the retention of fsu's only. Kathuria (1973) extended this study for two-stage sampling design on the occassion when partial retention is used either at first or at second or at both the stages for all the occasions. He considered two-phase cum two-stage sampling for two occasions. He also extended the method of Goswami and Sukhatme (1965), developed for double sampling with p auxiliary variates, for two stage sampling design over two occasions. All these results are essentially for the situation where sizes of fsu's are large.

Agarwal and Gupta (1975) studied the problem of multi-phase sampling in a two-stage design under general correlation pattern for two occasions and general replacement pattern when fsu's and ssu's both are partially retained. The minimum variance linear unbiased estimator and its variance were obtained where fsu sizes need not be large. Gupta and Agarwal (1975) also obtained the results for two occasions using the product method of estimation in a two-stage design where the fsu's are selected with ppswr. Agarwal (1977), Agarwal and Tikkiwal (1980, 1982) developed the theory of two stage sampling on successive occasions under less restrictive assumptions. The authors, further, extended the results to multi-stage sampling, and multi-phase cum multi-stage sampling on h (≥ 2) occasions. The authors also discussed minimum variance linear unbiased estimation procedure of population mean on different occasions for specified characters in specified T-classes.

EXERCISES

7.1. Consider the successive sampling scheme on two occasions. On the first occasion a sample s_1 and n_1 units is selected with SRSWOR from a finite population π of size N. On the second occasion a sub-sample s_1' of n_2' units is selected with SRSWOR from s_1 and an independent sample s_2' of n_2'' units is selected (such that $n_2 = n_2' + n_2''$) from the population $\pi - s_1$ by SRSWOR. Obtain an MVLUE in T_1-class of population mean for the second occasions μ_2 and that variance of the estimator. Further, if $n_1 = n_2 = n$, obtain optimum value of the fraction (n_2'/n) to be matched.

7.2 Consider the successive sampling scheme on two occasions, assuming $n_1 = n_2 = n$. On the first occasion a sample s_1 of n units is selected with SRSWOR from population π of size N. On the second occasion a sub-sample s_{2m} of m units is selected with SRSWOR from s_1 and an independent sample s_{2u} of u units is selected with SRSWOR from the population $\pi - s_1$ such that $n = u + m$. If the total cost of the survey is $C = C_0 + C_1 m + C_2 u$, obtain the optimum values of u and m so as to minimize the variance of $\hat{X}_2$, defined in Eq. (7.3.1) for h=2, for a fixed value of C, taking $n_2' = m$ and $n_2'' = u$. [Kulldroff (1963)]

7.3 Consider successive sampling scheme on two occasions. On the first occasion a sample s_1 of n units is selected with ppswr from a population π using x as a size variable ($p_i = x_i/x$, size measure of i). On the second occasion a sub-sample s_{2m} of m units is selected from s_1 with SRSWOR and an independent sample s_{2u} of u (= n – m) units is selected with ppswr from the population π. Show that following estimator $\hat{Y}_2$ is as unbiased estimator of population total t_2 for the second occasion and obtain its variance

$$\hat{Y}_2 = \phi\, t_{2m} + (1-\phi)\, t_{2u}, \qquad 0 < \phi < 1$$

where,

$$t_{2m} = \sum_{i \in s_{2m}} \frac{(y_{2j} - y_{1j})}{m\, p_j} + \sum_{j \in s_1} \frac{y_{1j}}{n p_j}$$

$$t_{2u} = \sum_{j \in s_{2u}} \frac{y_{2j}}{u\, p_j}$$

where, y_{ij} is the value of variable under study y on the j-th unit in the population in the i-th occasion (i = 1, 2 ; j = 1, . . . , N).

[Des Raj (1965)]

7.4 Consider successive sampling scheme on two occasions. On the first occasion a sample s_1 of n units is selected from population π by the Rao-Hartley-Cochran (1962) sampling procedure using $p_i = x_i/X$. On the second occasion a sub-sample s_{2m} of m units is selected from s_1 by SRSWOR scheme and an independent sample s_{2u} of u (= n – m) units is selected from the population π by using RHC- procedure, using the same p_i values. Show that the following estimator $\hat{Y}_2'$ is unbiased of population total t_2 for the second occasion and obtain its variance.

$$\hat{Y}_2' = \phi\, t_{2m} + (1-\phi)\, t_{2u}, \qquad 0 < \phi < 1$$

where,

$$t_{2m} = \sum_{j \in s_{2m}} \frac{(y_{2j} - y_{1j})\, n\, \pi_j}{m\, p_j} + \sum_{j \in s_1} \frac{y_{1j} \cdot \pi_j}{p_j},$$

$$t_{2u} = \sum_{j \in s_{2u}} \frac{y_{2j}\, \pi_j^*}{p_j}$$

and π_j, π_j^* denote the sums of p_t-values included in the groups of unit j while selecting s_1 and s_{2u} respectively by the RHC method.

[Ghangurde and Rao (1969)]

7.5 Consider successive sampling scheme on two occasions. On the first occasion a sample s_1 of size n is selected with SRSWOR from population π of size N. On the second occasion a sub-sample s_1' of n_2' units is selected from s_1 with SRSWOR and an independent sample s_2' of n_2'' units, such that $n_2' + n_2'' = n$, is selected from the population $\pi - s_2'$. Show that an unbiased estimator of μ_2 is given by

$$\hat{X}_2' = \theta\, t + (1-\theta)\; t_{2n_2'}$$

where, $$t = \left[(N - n_2')\; X_{2n_2''} + n_2'\, X_{2n_2'}\right] / N,$$

$$t_{2n_2'} = X_{2n_2'} + b\; (X_{1n} - X_{1n_2'}),$$

X_{1n} is the sample mean of n units on the first occasion.

b is the sample regression coefficient and θ is a suitable constant; other notations are as defined in Section 7.3. Further, with the optimum choice of θ show that,

$$V(\hat{X}_2') = \frac{(1-f)(1-\lambda f)\left\{\lambda\;(1-f) + (1-\lambda f)(1-\rho^2)\right\} S_2^2}{\lambda\;(1-f)\;(2-\lambda-\lambda f) + (1-\lambda)^2\;(1-\rho^2)\, n}$$

where, $f = n/N$, $\lambda = n_2'/n$, $\rho_{12} = \rho$. Hence show that optimum value of λ is

$$\hat{\lambda} = \frac{\sqrt{1-\rho^2}}{\left(1 - f + \sqrt{1-\rho^2}\right)}$$

and corresponding optimum value of variance is

$$V_{opt} = \frac{(1-f)\left(1+\sqrt{1-\rho^2}\right) S_2^2}{2n}$$

[Singh, R. (1972)]

7.6 Show that the estimator $\hat{X}_h$, for $h \geq 2$ given in Eq. (7.3.1) is MVLUE for μ_h is T_1-class, and also obtain its variance.

7.7 Consider the estimator $\hat{X}_h$ of the population means, μ_h on h-th occasion ($h \geq 2$), gives in Eq. (7.3.1). That is,

$$\hat{X}_h = (1-\phi_h)\; Z_h + \phi_n\; X_{h n_h''}$$

suppose the various regression and correlation coeffcients occuring in $\hat{X}_h$ are now estimated from the common units, then show that variance of $\hat{X}_h$ is given by

$$V(\hat{X}_h) = \frac{\sigma_h^2}{n_h''} E(\hat{\phi}_h)$$

$\hat{\phi}_h$ being the sample estimate of ϕ_h. This is so, provided n_t' common units on the t-th occasion forms a sub-sample of n_{t-1}'' new units on (t – 1) th occasion for t > 2. If this condition is not satisfied and if the various correlation and regression coefficients are still estimated from the common units then show that $\hat{X}_h$ will be consistent and asymptotically unbiased estimator of μ_h and its mean sequence error V will now satisfy the inequality

$$\phi_h \frac{\sigma_h^2}{n_h''} \le V \le \frac{\sigma_h^2}{n_h''} E(\hat{\phi}_h)$$

Further show that a consistent asymptotically unbiased estimator $\hat{V}$ of the variance or the mean square error of $\hat{X}_h$, as the case may be, is given by

$$\hat{V} = \hat{\phi}_h \frac{\hat{\sigma}_h^2}{n_h''}$$

where $\hat{\sigma}_h^2$ is the usual unbiased estimator of σ_h^2 based on n_h units.

[Tikkiwal, B.D. (1956), Gupta and Tikkiwal, G.C. (1998)]

7.8 Let y denote the characteristic under study and x the auxiliary variable. Under these notations consider the double sampling regression estimator

$$\overline{Y}_{ld} = \overline{Y}_n + \hat{\beta}(\overline{X}_{n'} - \overline{X}_n)$$

where n is a sub-sample of a larger sample of size n' ; both the sub-sample and the larger sample are selected with equal probability without replacement and (y_i, x_i) follows a bivariate normal distribution with means (μ_y, μ_x), variances (σ_y^2, σ_x^2) and correlation coefficient ρ. Show that $\overline{Y}_{ld}$ is an unbiased estimator of μ_y and that its variance is given by

$$V(\overline{Y}_{ld}) = \sigma_y^2 (1-\rho^2)\left(\frac{1}{n}-\frac{1}{n'}\right)\left(1+\frac{1}{n-3}\right) + \frac{\sigma_y^2}{n'}$$

Also show that an unbiased estimate of the variance is given by

$$\hat{V}(\bar{y}_{ld}) = \left(\frac{1}{n} - \frac{1}{n'}\right)\frac{Q}{n-3} + \frac{s_y^2}{n'}$$

where, $$Q = \sum_{i=1}^{n} \left[(y_i - \bar{y}_n) - \hat{\beta}\,(x_i - \bar{X}_n)\right]^2$$

and, $$\hat{\beta} = \frac{s_{xy}}{s_x^2}$$

$$s_{xy} = \frac{1}{n-1}\sum_{i=1}^{n}(x_i - \bar{X}_n)(y_i - \bar{Y}_n), \quad s_x^2 = \frac{1}{n-1}\sum_{i=1}^{n}(x_i - \bar{X}_n)^2$$

[Tikkiwal, B.D. (1960)]

REFERENCES

Abraham, T.P., Khosla, R.K. and Kathuria, O.P. (1969). Some investigations on the use of successive sampling in pest and disease surveys. J. Ind. Soc. Ag. Statist., **21**, 43-57.

Agarwal, C.L. (1977). Multiphase cum multistage sampling on successive occasions. Ph.D. thesis, University of Rajasthan, Jaipur, India under the supervision of Prof. B.D. Tikkiwal.

Agarwal, C.L. and Gupta, P.C. (1975). Multipurpose surveys on successive occasions in a two-stage design. Abstract, J. Ind. Soc. Ag. Statist., **27** (1), 114.

Agarwal, C.L. and Tikkiwal, B.D. (1980). Two stage sampling on successive occasions. Sankhya, **42**, Ser. C, 31-44.

Agarwal, C.L. and Tikkiwal, B.D. (1982). A note on minimum variance linear unbiased estimators in multistage sampling design on successive occasions. J. Ind. Soc. Ag. Statist., **34** (2), 76-86.

Avadhani, M.S. and Sukhatme, B.V. (1970). A Comparison of two sampling procedures with an application to successive sampling. J. Roy. Statist. Soc., Ser. C, 251-259.

Avadhani, M.S. and Sukhatme, B.V. (1972 a). Estimation in sampling on two successive occasions. Statistica, Neerlandica, **26** (2).

Avadhani, M.S. and Sukhatme, B.V (1972 b). Sampling on several successive occasions with equal and unequal probabilities and without replacement. Austr. J. Statist., **14** (2), 109.

Avadhani, M.S. and Srivastava, A.K. (1972). A comparison of Midzuno-Sen scheme with pps sapling without replacement and its application to successive sampling. Ann. Inst. Statist. Maths., **24**(1), 153-164.

Basu, D. (1958). On sampling with and without replacement. Sankhya, **20**, 287-294.

Bhargava, N.K. (1975). On some aspects of sampling with special reference to T-classes of linear estimators. Ph.D. Thesis, University of Rajasthan, Jaipur, under the supervision of Prof. B.D. Tikkiwal.

Bhargava, N.K. (1978). On some applications of the technique of combined unordering. Sankhya, **10** (1), Ser. C., 74-83.

Bhargava, N.K. and Tikkiwal, B.D. (1978). On the axioms of sample formation with their relevance to T-classes of linear estimators.

Sankhya, Ser. C, Part- 1, 9-18. Abstract of this paper appeared earlier in J. Ind. Soc. Ag. Statist. (1973) **25,** 220.

Bhargava, N.K. and Tikkiwal, B.D. (1974). On minimum variance unbiased linear estimation in T-classes. Abstract, Inst. Math. Statist. Bulletin.

Bhargava, N.K., Tikkiwal, B.D. and Tikkiwal, G.C. (1991). On minimum variance linear unbiased estimation in T-classes. J. Ind. Soc. Ag. Statist. **43** (1), 59-70.

Cochran, W.G. (1977). Sampling techniques. 3rd Ed., John Wiley and Sons, New York.

Des Raj (1956). Ratio estimators in sampling with varying probabilities without replacement. J. Amer. Statist. Assoc., **51**, 269-284.

Des Raj (1958). Sampling Theory. McGraw Hill Book Company.

Des Raj (1965). On sampling over two occasions with probability proportional to size. Ann. Math. Statist., **36**, 327-330.

Des Raj and Khamis, S.H. (1958). Some remarks on Sampling with replacement. Ann. Math. Statist., **29**, 550-557.

Durbin, J. (1953). Some results in sampling theory when the units are selected with unequal probabilities. J. Roy. Statist. Soc., **15**, Ser. B, 262-269.

Eciomovic, J.P. (1956). Three-stage sampling with varying probabilities of selection. J. Ind. Soc. Ag. Statist., **8**, 14-44.

Fellegi, I.P. (1963). Sampling with varying probabilities without replacement; Rotating and non-rotating samples. J. Amer. Statist. Assoc., **58**, 183-201.

Feller, W. (1972). An introduction to probability theory and its applications. Vol. 1, 3rd Ed., Wiley Eastern Print.

Ghangurde, P.D. and Rao, J.N.K. (1969). Some results on sampling over two occasions. Sankhya, Ser. A, **31**, 463-472.

Godambe, V.P. (1955). A unified theory of sampling from finite populations. J. Roy. Statist. Soc., Ser. B, **7**, 269-272.

Godambe, V.P. (1965). A review of the contribution towards a unified theory of sampling from finite populations. Rev. Inter. Statist. Inst., **33**, 242-258.

Gupta, A.K. and Tikkiwal, G.C. (1998). On variance of the estimator of population mean on $h(\geq 2)$-th occasion. J. Statist. Res., **32**(2), 49-58.

Gupta, M.G. (1964). Some unpublished results on varying probabilities. Department of Statistics, University of Rajasthan, Jaipur.

Hansen, M.H. and Hurwitz, W.N. (1943). On the theory of sampling from finite populations. Ann. Math. Statist., **14**, 333-362.

Hansen, M.H. and Hurwitz, W.N. (1949). On the determination of optimum probabilities. Ann. Math. Statist., **20,** 426-432.

Hanurao, T.V. (1962). An existence theorem in sampling. Sankhya, Ser. A, **24** (3), 327-330.

Hartley, H.O. and Rao, J.N.K. (1962). Sampling with unequal probabilities and without replacement. Ann. Math. Statist., **33**, 350-374.

Horvitz, D.G. and Thompson, D.J. (1952). A generalization of sampling without replacement from a finite population. J. Amer. Statist. Assoc., **47**, 663-685.

Jayanti Chotai (1974). A note on Rao-Hartley-Cochran method for pps sampling over two occasions. Sankhya, Ser. C, **36,** 175-180.

Jessen, R.J. (1942). Statistical investigation of a sample survey for obtaining farm facts. Iowa Ag. Expt. Station Res. Bull. No. 304, Ames, Iowa.

Kathuria, O.P. (1973). On alternative replacement procedures in sampling on successive occasions with a two-stage design and on use of multi-auxiliary information in such designs. Ph.D. Thesis, IARS, New Delhi.

Kathuria, O.P. and Singh, D. (1971 a). Comparison of estimates in two-stage sampling on successive occasions. J. Ind. Soc. Ag. Statist., **23**, 31-51.

Kathuria, O.P. and Singh, D. (1971 b). Relative efficiencies of some alternative replacement procedures in two-stage sampling on successive occasions. J. Ind. Soc. Ag. Statist., **23** , 101-114.

Konijn, H.S. (1973). Statistical theory of sample survey design and analysis. North-Holland Publishing Company, Amesterdam-London.

Koop, J.C. (1961). Contributions to the general theory of sampling from finite populations without replacement and with unequal probabilities. Mimeo Series No. 296, Institute of Statistics, North Carolina, USA based on his Ph.D. thesis (1957) at the Institute itself.

Koop, J.C. (1963). On the axioms of sample formation and their bearing on the construction of linear estimators in sampling theory for finite population. Metrika, **7** (2 and 3), 8-114, 165-204.

Kulldroff, G. (1963). Some problem of optimum allocation for sampling on two occasions. Rev. Int. Statist. Inst., **31**, 24-57.

Lahiri, D.B. (1951). A method of sample selection providing unbiased ratio estimates. Bull. Int. Statist. Inst., **3**, 133-40.

Lanke, J. (1974). On non-negative variance estimators in survey sampling. Sankhya, Ser. C, **36**, 33-42.

Lanke, J. (1975). Some contributions to the theory of survey sampling. Ph.D. Thesis, Department of Mathematical Statistics, Box 725 S-220007, Lund, Sweden.

Mahalanobis, P.C. (1939). Sample survey of the acreage of jute in Bengal. Sankhya, **4,** 511-531.

Midzuno, H. (1950). An outline of the theory of sampling systems. Ann. Inst. Statist. Math. (Japan), **1,** 149-156.

Murthy, M.N. (1957). Ordered and unordered estimators in sampling without replacement. Sankhya, **18**, 379-390.

Narain, R.D. (1951). On sampling without replacement with varying probabilities. J. Ind. Soc. Ag. Statist., **3,** 169-174.

Pathak, P.K. (1964). Sufficiency in sampling theory. Ann. Math. Statist., **35,** 795-808.

Pathak, P.K. (1965). Recent advances in sampling theory. Publication No. SM 65-4, Research and Training School, Indian Statistical Institute, Calcutta.

Pathak, P.K. and Rao, T.J. (1967). Inadmissibility of customary estimators in sampling over two occasions. Sankhya, **29**, Ser. A, 49-54.

Patterson, H.D. (1950). Sampling on successive occasions with partial replacement of units. J. Roy. Statist. Soc., B, **12**, 241-55.

Prabhu Ajgaonkar, S.G. (1962). Some aspects of successive sampling. Ph.D. Thesis, Karnatak University, Dharwar, India, under the supervision of Prof. B.D. Tikkiwal.

Prabhu Ajgaonkar, S.G. (1967). On unordering best estimator in Horvitz and Thompson's T_1 – class of linear estimators. Sankhya, **89**, Ser. B, 209-212.

Prabhu Ajgaonkar, S.G. (1967 a). Unbiased estimators of the variance of the Narain, Horvitz and Thompson estimator. Sankhya, **29,** Ser. A, 55-60.

Prabhu Ajgaonkar, S.G. and Tikkiwal, B.D. (1961). On Horvitz and Thompson's T-class estimators. Abstract, Ann. Math. Statist., **32**, 923.

Purakam, N. and Koop, J.C. (1966). Multistage sampling on successive occasions where first-stage units are drawn with unequal probabilities and with replacement. Mimeographed Series No. 472 of the North Carolina State University, Raleigh, N.C.

Ramakrishnan, M.K. (1970). Optimum estimators and strategies in survey sampling. Ph.D. thesis submitted to the Indian Statistical Institute, Calcutta.

Rao, J.N.K. (1961). On the estimate of variance in unequal probability sampling. Ann. Inst. Statist. Math., **13** (1), 57-60.

Rao, J.N.K. (1962). On sampling with varying probabilities and with replacement in sub-sampling designs. J. Ind. Soc. Ag. Statist., **13**, 211-218.

Rao, J.N.K. (1963). On three simple procedures of unequal probability sampling without replacement. J. Amer. Statist. Assoc., **59**, 202-215.

Rao, J.N.K. (1965). On two simple procedures of unequal probability sampling without replacement. J. Ind. Soc. Ag. Statist., **3**, 173-180.

Rao, J.N.K., Hartley, H.O. and Cochran, W.G. (1962). On a simple procedure of unequal probability sampling without replacement. J. Roy. Statist. Soc. **24,** Ser. B, 482-491.

Sen, A.R. (1953). On the estimate of the variance in sampling with varying probabilities. J. Ind. Soc. Ag. Statist., **5**, 119-127.

Sen, A.R. (1953). Recent advances in sampling with varying probabilities. Calcutta Statist. Assoc. Bull., **5**, 1-15.

Seth, G.R. (1966). On estimators of variance of estimate of population total in varying probabilities. J. Ind. Soc. Ag. Statist., **18**, 52-56.

Seth, G.R. and Rao, J.N.K. (1964). On the comparison between simple random sampling with and without replacement. Sankhya, **26,** Ser. A, 85-86.

Singh, D. (1968). Estimation in successive sampling using a multi-state design. J. Amer. Statist. Assoc., **63**, 99-112.

Singh, D. and Kathuria, O.P. (1969). On two-stage successive sampling. Austr. J. Statist.,**11**, 59-66.

Singh, R. (1972). On Pathak and Rao's estimate on pps with replacement sampling on two occasions. Sankhya, **34,** Ser. A, 301-303.

Srivastava, A.K. and Shivtar Singh (1974). A note on two-stage successive sampling. J. Ind. Soc. Ag. Statist., **26** (1), 57-64.

Sukhatme, P.V. and Narain, R.D. (1962). Sampling with replacement, J. Ind. Soc. Ag. Statist., **4**, 42-49.

Sukhatme, P.V. and Sukhatme, B.V. (1970). Sampling theory of surveys with applications. Ind. Soc. Ag. Statist., New Delhi, India and the Iowa State College Press, Ames, Iowa, USA. Revised edition (1984) by the two authors alongwith Dr. (Mrs.) S. Sukhatme and Dr. C. Asok.

Tikkiwal, B.D. (1950). Estimation by successive sampling. A paper presented before the 4th Annual Meeting of the Ind. Soc. Ag. Statist., New Delhi.

Tikkiwal, B.D. (1951). Theory of successive sampling. Thesis submitted towards partial fulfiment of requirements for Diploma in Agricultural Statistics at present Indian Agricultural Statistics Research Institute, New Delhi.

Tikkiwal, B.D. (1953). Optimum allocation in successive sampling. J. Ind. Soc. Ag. Statist., **5,** 100-102.

Tikkiwal, B.D. (1955). Multiphase sampling on successive occasions. Ph.D. Thesis, North Carolina State College, Raleigh.

Tikkiwal, B.D. (1956). A further contribution to the theory of univariate sampling on successive occasions. J. Ind. Soc. Ag. Statist., **8**, 84-90.

Tikkiwal, B.D. (1958). Theory of successive two-stage sampling. Abstract, Ann. Math. Statist., **29,** 1291.

Tikkiwal, B.D. (1960). On the theory of classical regression and double sampling estimator. J. Roy. Statist. Soc., **22,** Ser. B, 131-138.

Tikkiwal, B.D. (1960). A course of lectures on T-classes of linear estimators. Karnataka University, Dharwar, India.

Tikkiwal, B.D. (1965). Some aspects of sampling with varying probabilities from a finite population. Rajasthan University Studies in Statistics (Science Series), Jaipur, India, 1-21.

Tikkiwal, B.D. (1965). The theory of two stage sampling on successive occasions. J. Ind. Soc. Ag. Statist., **3** (2), 125-135.

Tikkiwal, B.D. (1966). On T_2 and T_3 – classes of estimators in sampling with varying probabilities and without replacement. J. Ind. Soc. Ag. Statist., **18,** 62-72.

Tikkiwal, B.D. (1967). Theory of multiphase sampling from a finite or an infinite population on successive occasions. Rev. Int. Statist. Inst., **35,** 247-263.

Tikkiwal, B.D. (1972). On unordering of estimators in sampling with or without replacement and its impact on T-classes of estimators. Abstract, I.M.S. Bulletin.

Tikkiwal, B.D. (1979). Successive sampling – A review. An invited paper presented as a principal speaker at the 42nd Biennial Session of International Statistical Institute, Manila.

Tikkiwal, B.D. (1985). The theory of T-classes of linear estimators, its generic value and its impact on teaching. Paper presented at the 45th Biennial Session of the International Statistical Institute, The Netherlands.

Tikkiwal, B.D. (1985). Some aspects of elements of theory of sample surveys. Lecture Notes, Department of Statistics, University of Rajasthan, Jaipur, India.

Tikkiwal, B.D. and Tikkiwal, G.C. (1979). Theoretical foundations of inference from finite population and data analysis. A paper in the proceedings of International Conference in Statistics at Tokyo, 85-88.

Tikkiwal, G.C. (1988). A note on T-classes of linear estimators. Ganita Sandesh, **2** (1), 32-38.

Tikkiwal, G.C. and Gupta, A.K. (1991). Estimation of population mean under successive sampling scheme when various weights and regression coefficients are unknown. Biom. J. **33** (5), 1-10.

Tikkiwal, G.C. and Jain, V. (2001). On the technique of combined unordering – an alternative approach to sufficiency in survey sampling. Aligarh J. Statist., **21**, 59-69.

Wilder, R.L. (1952). Introduction to foundations of mathematics, John Wiley and Sons, Inc., New York, London, Sydney.

Yates, F. (1949). Sampling methods for census surveys. Charles Griffin and Co., London.

Yates, F. and Grundy, P.M. (1953). Selection without replacement from within strata with probability proportional to size. J. Roy. Statist. Soc., **15,** Ser. B, 253-261.

SUBJECT INDEX